Dedicated to
Prof. Dr. Ludwig Armbruster
(7. 9. 1886 — 4. 6. 1978)
in grateful remembrance

Breeding the Honeybee

A Contribution to the Science of Beebreeding

By Brother Adam

NORTHERN BEE BOOKS
Mytholmroyd : Hebden Bridge

This first English edition published by Northern Bee Books, Scout Bottom Farm, Mytholmroyd, Hebden Bridge, West Yorkshire. March 1987.

The translation of this volume was personally supervised by Brother Adam from the original German.

'Zuchtung der Honigbiene' was first published in 1982.

British Library Cataloguing in Publication Data

Adam, *Brother*
Breeding the honeybee: a contribution to the science of bee-breeding.
1. Bee culture—Europe
I. Title   II. Züchtung der Honigbiene.
English
638'.1   SF531.E9

ISBN 978-1-912271-73-3

*Brother Adam
born August 3rd, 1898*

Contents

Preface	1
The Honeybee — then and now	5
Nature as a Breeder	6
Results of Recent Breeding Endeavours	7

Part I
The Theory of Breeding

The Bee as a Member of a Social Unit	11
The Way of Life of Bees and their Adaptability to Environment	12
Why Bee Breeding is an Exceptional Case	14
The Effects of Parthenogenesis	15
The Pedigree of the Honeybee	16
The Significance of Multiple Mating	19
The Advantages and Disadvantages of Inbreeding	22
Mendel's Laws of Heredity	23
Racial Purity in the Light of these Laws	31
The Application of Mendelism to the Honeybee	32
Chromosomes and Reduction Division	36
The Reciprocal Influence of the Genes	38
Results secured in the crosses indicated	40
Polymeral Transmission	41
Linked Characteristics	42
The Limitations Imposed by Heredity	42
Mutations	43
Synthesising New Combinations	45
The Limits Imposed on Us	46
Sex Determination	47
A Word of Warning about Mere Theory	49

Part II
Practical Possibilities in Breeding

Preliminary Remarks	53
The Aims of Breeding	55
The Economic Goals in Breeding	55
The Primary Qualities for Performance	56
1. Fecundity	56
2. Industry or Foraging Zeal	57
3. Resistance to Disease	57
4. Disinclination to Swarm	58
Secondary Qualities	58
1. Longevity	59
2. Wing-power	59

3. Keen Sense of Smell	59
4. Instinct for Defence	60
5. Hardiness and Ability to Winter	60
6. Spring Development	60
7. Thrift	61
8. Instinct for Self Provisioning	61
9. Arrangement of the Honey Stores	61
10. Wax Production and Comb Building	62
11. Gathering of Pollen	62
12. Tongue — reach	62
Qualities which Influence Management	63
1. Good Temper	63
2. Calm Behaviour	63
3. Disinclination to Propolise	64
4. Brace Comb	64
5. Cleanliness	64
6. Honey Cappings	64
7. Sense of Orientation	65
Breeding as a means of Combating Disease	66
Resistance and Immunity	67
Contradictory Reports	67
Diseases of the Adult Bee	68
Acarine	68
Nosema	71
Paralysis	72
Bacterial Septicaemia	73
Diseases of the Brood	74
American Foulbrood	74
European Foulbrood	75
Sac Brood	75
Chalkbrood	76
Anomalies of the Brood	76
Summary	76
Evaluation of Performance	77
Breeding Procedures	79
Nature's Method of Breeding	80
Pure Breeding	80
Line Breeding	81
Cross Breeding	81
Combination Breeding	86
Development of New Combinations	87
Intensive Selection	88
Some Results in Combination Breeding	89
Multiple Hybrids	90

Part III
An Evaluation of the Breeding Possibilities of the Different Races of Honeybees

Introduction	93
Nature's Aims in Breeding	93
Acclimatisation	93
Environment	94
Standards of Reference	95

Biometric Findings	95
My Searches	96
The Essential Characteristics of the Races of Bees	96
Ligustica	96
Carnica	98
Sub-varieties of the Carnica	99
Cecropia	99
Caucasica	100
Anatolica	102
Fasciata Race Group	104
Fasciata	104
Syriaca	105
Cypria	105
Adami	107
Intermissa Race Group	108
Intermissa	108
Sub-varieties of the Intermissa	109
Northern European and North Asia Sub-varieties	110
Apis Mellifera Major Nova	111
Sahariensis	111
Results of the Evaluations in Relation to the Buckfast Strain	113
The Genetic Resources	114
Conclusion	116
Glossary	117

Preface

In the course of the years I have often been asked to record in writing my findings and views on breeding the honeybee. The requests were based on the fact that there is very little information available on this subject in our literature. A great deal has been written on the different ways of raising queens but almost next to nothing on breeding the honeybee on a genetic basis, apart from the publications of Prof. Dr. L. Armbruster. However, these appeared in German and were never translated. In fact his writings fell on stony ground in his own country, for the simple reason that he was too far ahead of his time.

My own findings cover a period close on seventy years of unremitting effort. They also embrace a first-hand knowledge of virtually all the races of the honeybee. This is an essential prerequisite to any serious attempt at an improvement of the honeybee on a genetic basis.

In 1910 F. W. Sladen of Ripple Court, Dover published a report in the British Bee Journal on an attempt he carried out in raising a new combination from a cross between the Old English native bee and a golden Italian strain developed in North America. These efforts of his were based on Mendel's discoveries. Sladen was therefore the first person to attempt a task of this kind, for Mendel's findings were re-discovered less than ten years previously. About the same time S. Simmins of Heathfield, Sussex endeavoured to produce a cross of outstanding economic value. He conclusively demonstrated the potentialities open to beekeeping in this direction. Indeed the results he secured had a far-reaching bearing on my own efforts.

Professor Dr. L. Armbruster's BIENENZUECHTUNGSKUNDE appeared in 1919. By a fortunate chance I obtained a copy in 1920. This book revealed to me a world of new possibilities. However, at that time nobody had an inkling of multiple matings so many of Armbruster's conclusions were based on false premises. On the other hand, his interpretations of Mendelian heredity in the light of parthenogenesis still hold good. Also his graphic illustrations and diagrams, explaining discoveries and conclusions, are superb. I am therefore including a number of them in this book.

Prof. Dr. L. Armbruster extended to me every possible help at all times. In recognition I am dedicating this book to his memory.

Apart from Armbruster I also owe a great debt of gratitude to a number of people who supported my efforts at one time or another, whose names I cannot enumerate here. I am likewise deeply grateful to those who actively assisted in the preparation and publication of this book — foremost, of course to the Right Rev. Abbot Leo Smith who undertook the translation from the German.

As the subtitle indicates, this book is merely a contribution to the science of breeding the honeybee on a genetic basis. However, I trust it will in some small measure fill a deficiency in our apicultural literature and also throw some light on the potentialities awaiting us. Nowadays 'genetic engineering' is receiving much attention. Cross breeding — the most practical form of genetic engineering — is clearly the only way open to us by which we can realize the potentialities the honeybee can offer.

Spring 1985   Br. Adam

Introduction

The Honeybee — then and now

It is now generally accepted that the earth has been in existence for some four to five thousand million years. The oldest honeybee has been preserved for us in amber in the Baltic, where it lived more than fifty million years ago. The fossil remains of the honeybee which have up to now been found in Europe date from a time previous to the Ice Age, that is from the Tertiary Period when the climatic conditions in Europe were similar to those prevalent in India today. The best known fossil remains of bees come from Randeck and Boettingen in south-west Germany, and also from Rott in the Siebengebirge. Those from Randeck and Boettingen belong to the early Miocene, those from Rott to the later Miocene Period. Apart from the remains which have come down to us in amber, the three forms which have been found in the Siebengebirge are the oldest primitive forms of the honeybee known to us. They date from some twenty-five million years ago.

All these primitive forms differ from one another in size and in a number of morphological characteristics. They all have a close resemblance to the *Apis mellifera*, although they are not identical with it. The *Sinapis dormitans* from Rott however is so close to our honeybee of today as to be easily mistaken for it.

None of these kinds of bees survived the Ice Age, which lasted for over a million years, its final retreat beginning some 30,000 years ago and ending about 10,000 years ago. During the various periods of the Ice Age it was impossible for the honeybee to exist in the greater part of Europe. The huge ice cap stretched from the North Pole southwards to a line drawn from the mouth of the Severn in England across to Kiev in Russia and then on further east. During this Ice Age the honeybee had only three places of refuge — the Iberian, Italian and Balkan peninsulas. In the areas north of the Pyrenees and the Alps as far as the edge of the ice cap there was but endless Tundra.

The bees of the Italian peninsula, the Ligustica, were probably confined to their native habitat, as the Alps formed an insuperable barrier to any form of migration northward. After the Ice Age the bees of the Balkans were able to spread north to the Eastern Alps and also north-east to the borders of Russia, where further progress was impeded not by the mountain ranges but by the treeless steppes. Hence at the end of the Ice Age the repopulation of the other European regions could be effected only by bees from the Iberian peninsula by way of the coastal gaps at either end of the Pyrenees rendering an unhindered migration northwards possible. These bees were a variation of the North-African *Apis m. intermissa* from which the West European races developed and formed the different sub-species.

Of the original forms of the honeybee which had their native habitat in northern Europe before the Ice Age none were left. True, the fossil remains provide us with information on the morphological side, but naturally we lack any means of ascertaining their physiological characteristics. Yet the almost unbelievable stability of the morphological features of these primitive forms and that of *Apis mellifera* of today, which have persisted throughout this length of time, provides us with a number of basic starting points which from the genetic point of view we cannot disregard. Although there is no evidence of any progressive evolution in the honeybee during these millions of years, there is clear indication of a ruthless natural selection.

Of the innumerable primitive forms of the honeybee only four have survived:
Apis mellifera
Apis cerana (Indian bee)
Apis dorsata (Giant bee)
Apis florea (Dwarf bee).
None of these kinds seems to owe its origin to the primitive bees whose fossil remains are known to us. The only possible exception is the *Sinapis dormitans* from Rott, which was mentioned above. This has a close resemblance to our *Apis mellifera*. Today *Apis mellifera* has spread throughout the entire world, except for south-east Asia which is the preserve of *Apis cerana*, *dorsata* and *florea*. These three kinds of honeybees are however of no real economic or breeding value. They cannot be crossed either among themselves or with the *Apis mellifera*.

Nature as a Breeder

From the outset the breeding and survival of the honeybee was left to the whim of Nature. Nature's aim in breeding is limited exclusively to the preservation and dissemination of a species and her sole means of doing this is a ruthless selection. Whatever could not adapt itself to a given environment was without exception left to its doom. The one aim was the survival of the most adaptable and the fittest. Although Nature has bequeathed to us only a small number of different types of honeybee, she has on the other hand provided us with a great number of different geographical races, ecotypes of immense value for breeding purposes. Yet true to her principles, Nature never breeds an ideal or 'perfect' bee, one which would answer all demands of the modern beekeeper. The realization of this ideal has been left by Nature to the progressive, purposeful bee breeder of today.

Results of Recent Breeding Endeavours

Bee breeding by up-to-date methods has hardly begun. Dr. Ulrich Kramer, a Swiss, provided in 1898 the initial impulse. However at that time Mendel's laws of genetics were hardly known and modern beekeeping was still in its infancy, at a stage of its development which has not even yet been completed. Before the introduction of the moveable frame in 1850, the whole life-process of a colony of bees was hidden behind a veil of mystery, and consequently any attempt at influencing beekeeping by breeding was impossible.

Again, bees which have been preserved by Nature for millions of years can survive in modern conditions and be an economic proposition without the beekeeper having to resort to breeding. There will be no notably high averages of honey per colony gained with a minimal expenditure of labour, but such beekeeping can be profitable. For these reasons then the experiments in bee breeding and the efforts made to produce a really profitable bee have till now produced very meagre results.

Apis dormitans
A petrified honeybee from about 25 million years ago that closely resembles our *Apis mellifera* of today.
(From "Archiv fuer Bienenkunde")

Up to the present time it is in the German-speaking countries that most work has been done in the field of bee breeding. It began in Switzerland with the 'Nigra', and for some thirty years the breeding experiments with this strain enjoyed a popularity not only in Switzerland but beyond its frontiers. Today however the 'Nigra' belongs to the pages of history. From about 1950 its place has been taken by the Carnica which today runs as the favourite, at least in Central Europe. Yet in spite of all that has been claimed for it, from the point of view of profitable bee-keeping it is difficult to see that any worthwhile progress has been made in breeding the Carnica as it is today. Her predecessor, the original Carniolan bee, had characteristics which were rightly highly esteemed, but these are lacking in the modern Carnica.

In the past fifty years the Ligustica has fared little better. True, we have a bee which as far as colour is concerned is more uniform and attractive, is good tempered and prolific, but short-lived and extremely unthrifty. She is really of value only where there are heavy flows and especially where the climatic conditions are favourable. To give an objective estimate from the strictly economic point of view of the breeding experiments made up to date one is forced to say that they have been only moderately successful.

The causes of this lack of progress are indeed numerous. Admittedly breeding the honeybee poses problems which are not encountered in the world of animals and plants. In fact from many points of view breeders in these fields have a comparatively easy task. On the other hand bee breeders have advantages which are denied to animal and plant breeders. Be that as it may, comparisons drawn from experiments in other realms of breeding have only very limited value when applied to the honeybee.

The real causes of the failure in bee breeding work up to the present have been the lack of a definite aim and the unrealistic, amateurish methods employed. In the following chapters we will try to set out what breeding the honeybee entails.

Part I
The Theory of Breeding

The Bee as a Member of a Social Unit

In general animal and plant breeding deals with individuals which are selected for the purpose of developing some favourable trait which they possess. In both these spheres of breeding the breeder has at his disposal an exact knowledge of the animal and plant structures, as well as those different desirable and undesirable characteristics which every individual living thing possesses. In other words, selection is concerned with definite individuals whose pedigree and whose 'personality' from the breeding point of view is known to the breeder, at least in general terms.

In the case of the honeybee however the breeder is confronted not with isolated individuals but with a society, or to put it more scientifically, with a superorganism, an extraordinarily well-regulated and well-ordered system, and a structure whose individual parts operate in perfect harmony. Moreover this social organism is in itself immortal, by which I mean that a colony of bees never dies of old age but only because of some misadventure due to external causes.

The first essential point then is that in breeding the honeybee we are dealing not with isolated individuals but with a society. It is however a special kind of society; it is a family with a mother, an indeterminate number of fathers (who have long been dead), a very large number of daughters and a limited number of males. Since the mother mated with an unknown number of drones (ten on the average) each colony contains an indeterminate number of groups of half-sisters. Each group has a common mother but fathers of different ancestry and hence diverse hereditary characteristics. The half- and full-sisters in each group and their respective influence on making the colony into a unity will be different in each case. Hence a colony of bees is an agglomeration of groups, each with its own definite set of hereditary dispositions, and it is these taken together which bestow on each community of bees its actual peculiar set of characteristics.

In the case of the honeybee we are dealing not with isolated individuals as in animal and plant breeding, but always with a community, one which moreover consists of groups of half-sisters of varying genetic dispositions, a fact we must always bear in mind.

The Way of Life of Bees and their Adaptability to Environment

In all forms of life environment plays a definite role in the development of the hereditary characteristics of the individual. A mere glance at animal and plant life is sufficient to prove this point. However much they may look alike outwardly, no two individuals of the same family or stock are absolutely identical. Even with self-propagating plants or with animals which although reproduced by cellular fission are absolutely identical genetically, environment has an essential influence on their growth and development. Any individual differences however occur only within the very definite limits set by their hereditary characteristics.

How does the honeybee fit into this general pattern? We know that the whole process of development from the egg to the mature bee, whether queen, worker or drone takes place in a milieu which is subject to only minimal fluctuations of temperature and humidity. The same holds good for food supplies, for the bees have the ability of adapting in a number of ways the size of the brood area and the rate of consumption of food to the amount of stores they have at hand. A complete lack of food means the death of the whole colony.

Into the world outside the colony, queens and drones normally venture only for the mating flight, and then only when weather conditions are favourable. It is only the worker bee which has at times to contend with an unfavourable environment. But even here there is no question of any quality acquired and transmitted. For the short span of life of the worker bee, occurring as it does at different times of the year, precludes any possibility of a permanent genetic adaptation to a particular local environment.

However in spite of thus being so unusually isolated from and independent of its environment, the honeybee is able to adapt itself (though this is not acclimatization) to conditions in the sense of a survival of the fittest. But this kind of adaptability is acquired only after a very long period of time and cannot adjust itself within a short space of time to any new conditions. An example of this is provided by an extremely prolific bee which turns all its stores into brood and then is brought to the point of starvation during an interruption of the honeyflow in the middle of summer. In such cases the result is an eventual disappearance of the strain, whereas in cases of less glaring defects, there is a continual weeding out of unsuitable individuals. We owe a debt of gratitude to Nature for this spontaneous natural selection. She has provided us in the course of thousands of years with those valuable ecotypes for breeding which we find in remote valleys and mountainous districts in addition to the geographically distinct races. All this material is now at the disposal of the progressive bee breeder.

It is widely assumed that a bee which has for a long time existed in certain surroundings and completely adapted itself to the prevailing conditions must of necessity be the most suitable bee for that region from the point of view of successful beekeeping. It is true that such a completely adapted bee can manage to survive in the worst of seasons. But Nature never breeds for performance but only to preserve a particular type. Hence as experience has conclusively proved, another race of bees brought in from a totally different environment can produce results which on average far surpass those of the native bee. In fact the introduction of different races and types of bees has many real advantages. For instance the Cyprian bee is able to winter in England much better than could the old native bee; we have never lost a colony of Cyprian bees due to the climate even in the wettest of winters and coldest of springs. And even in such adverse conditions when the time is at hand for the spring build-up they are always in the lead.

Bees then do adapt themselves over long periods of time to their environment, but they do not become acclimatized as was often generally but erroneously assumed. At one time hopes were raised that bees from the Middle East would become acclimatized to conditions in Central Europe and could be used for ordinary beekeeping purposes. The grounds for such an assumption were obviously false. Bees of such origin as the Middle East have indeed a value but it lies in another direction.

Why Bee Breeding is an Exceptional Case

There is a tendency among breeders of the honeybee to quote examples from and use the same terminology as in all other kinds of breeding. But in bee breeding we are confronted on all sides with an array of factors and difficulties which are simply unknown in animal breeding. As has already been mentioned, in bee breeding we are concerned not with isolated individuals but with a society made up of groups which possess very different hereditary characteristics and vary in strength during the course of the year. The breeding material, the queens and the drones, give us no indication, apart from the fertility of the queen, of the worthwhile factors they will bequeath to their progeny. The function of the parent bees is solely concerned with maintaining and developing the colony. The worker bees alone perform an economic function in the colony and alone manifest the characteristics with which we are concerned in breeding. Although they are females they reproduce only when compelled to by necessity, and then normally can only lay eggs that give rise to drones.

As I have said we come up against a whole host of problems in bee breeding which are unknown in other spheres of animal and plant breeding. Some of these problems are the parthenogenesis in the queens; multiple mating; mating with drones of unknown origin; the fact that every drone dies in the act of mating and hence cannot be used for further matings as is possible in other spheres of breeding. Although the honeybee is subject to the universally valid laws of genetics as laid down by Mendel, nevertheless it manifests exceptions and peculiarities of vital importance. As I shall point out later the laws which Mendel discovered about the segregation of the characteristics do not hold in the case of the honeybee exactly as in other forms of breeding. From the breeding point of view the honeybee occupies an exceptional place which has no counterpart in breeding in general.

The Effects of Parthenogenesis

Apart from the fact that we have at best only a limited control of the mating of the queens, it is parthenogenesis which is the real critical point in the breeding of the honeybee. Parthenogenesis not only nullifies the normal breeding processes, it also shatters all our preconceived notions and hypotheses concerning heredity. For due to parthenogenesis the drone has no father but only a mother. Moreover he loses his life in the act of mating and so ceases to be of service for purposes of further breeding. Hence there is no possibility in the honeybee of matings between father and daughter, mother and son, or brother and sister. At most the possibility is of mating half-brothers to half-sisters.

From this it follows that such crossings, as are regarded as indispensable in all other forms of breeding to intensify certain hereditary characteristics, are impossible in the honeybee. The situation is further complicated by the fact that due to parthenogenesis the millions of spermatozoa produced by a drone are all absolutely identical from the genetic point of view.

A result of this uniformity of genes in the drone means that there is a greater stability in the heredity of the honeybee than in other forms of life. Another consequence of this uniformity is that the honeybee is more susceptible to inbreeding. It is true that multiple mating acts as a counterbalance to this, but only partially. In our cross-breeding we obtain a segregation in the female offspring in the F1, in the drones only in the F2, but after that there is not the same pattern as in other types of breeding where there is no parthenogenesis. By crossing individuals of the F1 among themselves Mendel was able to obtain the classical segregation in the F2, from which arise the new combinations of genes which are then transmitted in a direct line. Such new combinations are possible in the case of the honeybee, but as has already been noted, only in a roundabout way because of the fact of parthenogenesis.

One point should be noted to clarify something that was said above. The millions of spermatozoa produced by a drone are all genetically identical. This does not however mean that the drone progeny of a queen are all uniform. In her male progeny a queen manifests the manifold nature of her own hereditary qualities as well as those of her parents, the grandparents of the drone.

The Pedigree of the Honeybee

Because of parthenogenesis the pedigree of the honeybee is essentially different from that of other forms of life. Normally among animals each individual has two parents, four grandparents, eight great-grandparents, sixteen great, great grandparents and so on, always double the number of its immediate ancestors. In the case of the honeybee the drone has one parent — his mother — two grandparents, three great-grandparents and only five great, great, grandparents. Queens and worker bees have two parents, but only three grandparents, five great-grandparents and eight great, great, grandparents, that is, only half the number according to the general rule. But the drone has not even half the usual number of great, great, grandparents. It is clear from this that the pedigree of the honeybee differs essentially from that of other forms of living beings. Not only is the pedigree of both sexes different as regards the number of ancestors, but the drone has neither father nor sons; he has only a grandfather and grandsons. In addition to this the pedigree of the honeybee becomes even more complicated because of multiple mating. But leaving that aside for the moment we can see with the help of a little arithmetic how the pedigree of the honeybee differs from that of a normal genealogy.

A normal genealogical table looks like this:

Breeder	Parents	Grandparents	Great-Grandparents	Great, Great, Grandparents
1	2	4	8	16

In queen bees and worker bees it is as follows:

| 1 | 2 | 3 | 5 | 8 |

or perhaps even more clearly when set out thus:

| 1 | 2 | 1+2 | 2+3 | 3+5 |

that is, each member is the sum of two previous ones.

The ancestors of the drone have likewise their own set of figures. The series of queen/workers and of drones both begin with the breeding mother, but the first member of the series in the drone line is always 1, whereas the former is always 2.

It is obvious that as far as the sex ratio is concerned the honeybee is quite an exception to the normal genealogical tree, where the series is always based on the power of 2. In the case of the honeybee the males are clearly in a minority as the following table shows:

	Parents	Grandparents	Great-Grandparents	Great, Great, Grandparents	etc.
Males	1	1	2	3	5
Females	1	2	3	5	8

The sex ratio therefore is:

	1–1	1–2	2–3	3–5	5–8
Females	50%	66.6%	60%	62.5%	61.53%

Thus some 61.8% in the pedigree of the bee are females.

Moreover in both sexes there is a progressive loss of the number of ancestors. In comparison with a normal genealogy the queen in the fifth generation has lost half and the drone more than two-thirds of the number of ancestors.

In view of this restriction and inequality of the number of ancestors the question arises as to the relative breeding worth of queen and drone. From what has been said above one is forced to conclude that in theory the dominant role must be given to the queen. This is in fact borne out in practice. The exact opposite occurs in the breeding of domestic animals where the father has the major role in determining the characteristics of his progeny. There are a number of reasons for this. Let us take as an example the breeding of cattle or of horses. The cow or mare can have direct influence on little more than twelve descendants at the most, whereas the bull or stallion can influence an almost indefinite number. In the case of the honeybee however the opposite is true. An individual drone plays a very modest part from the breeding point of view. And this part is even further reduced by multiple mating, since only very few of his descendants will ever themselves reproduce, only, that is, those which develop into queens and are mated. Even in the most favourable circumstances these will form only a very small part of his direct descendants. Moreover we can never say that this or that queen is the daughter of a particular drone, as is the case with domestic animals. Once a drone has successfully mated he can play no further part in breeding. Yet, in spite of the many disadvantages under which the drone labours, we must never underestimate his value in breeding.

The dominant influence exercised by the mother in breeding extends to all the characteristics of her descendants. Hence in comparison with the breeders of domestic animals bee breeders have a number of advantages. The breeding potential and range of selection are in the case of the honeybee virtually unlimited. Moreover we are able to carry out preliminary tests of performance without appreciable expense before we are committed to an extensive programme of breeding.

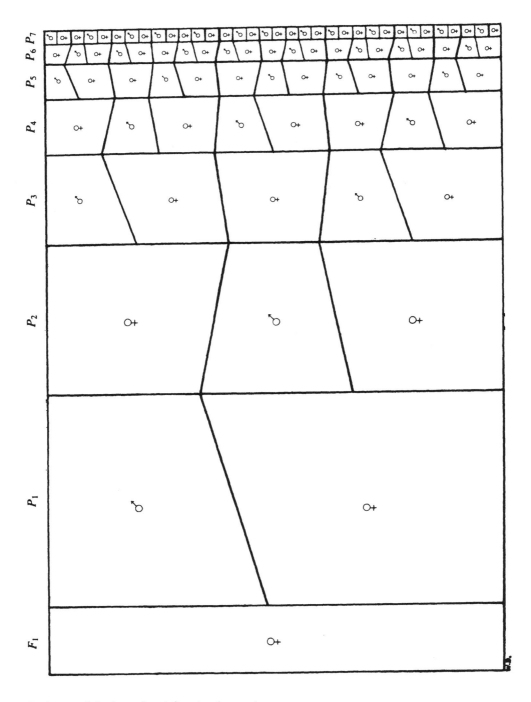

Pedigree of the honeybee (after Armbruster)

The Significance of Multiple Mating

At one time even the possibility of multiple mating was regarded as doubtful, both by scientists and practical beekeepers. Some 200 years ago Anton Janscha and Francois Huber noticed that occasionally a queen returned with the mating sign several times from mating flights. Both men thought that the reason for this was that previous matings had not been successful. This assumption which was confirmed by experience in the rest of the animal world was taken for granted until 1944. In that year Dr. W. C. Roberts of Baton Rouge, Louisiana, published his findings on 110 queens he had observed. More than half were seen several times with the mating sign. My own experiments which were conducted in May 1947 showed convincingly that multiple mating was far more common than had been previously accepted. It was still however thought that it was due to an ineffective mating and this notion was strengthened by the experiments with instrumental insemination which were just then proving to be a practical proposition. It was clear that a certain percentage of drones possessed greatly varying amounts of semen, and some had no sperm at all, and this in spite of the very best breeding. At the same time it was clearly shown that a completely sexually mature drone possessed almost double the amount of semen required for filling the spermatheca of a queen.

The publication of Dr. Robert's findings at once provoked discussion, since the possibility of genuine multiple mating posed crucial problems in connection with breeding. What was needed was a clear decisive rejection or confirmation of Dr. Robert's work. This was conclusively provided by the experiments made by Prof. Ruttner in the summer of 1953 and by the research he carried out in the following year on the island of Volcano off the coast of Italy. Experiments were made by controlling the flight of drones of the

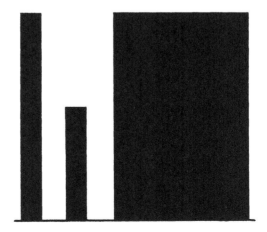

Diagram indicating the proportional amounts of semen: left, the quantity of semen produced by a single drone; middle column, the capacity of a spermatheca; right, the corresponding volume of semen recovered from the oviducts of a queen on her return from a mating flight.
(After Prof. Dr. Ruttner)

Carnica, European black bee and Cyprian on different days and then examining their progeny. It was clear that after each mating semen was transferred to the spermatheca of the queen. This had been called in doubt until then. At the same time the observations made on Volcano showed that more than half of the 114 queens tested had the mating sign more than once. Now came the great surprise. It was found that even on one and the same flight queens had mated with several drones. This was something quite new and had never before been even suspected. But Prof. Ruttner's examinations of queens under the miscroscope left no doubt that on their return from a mating flight they had a far greater amount of semen in their oviducts than even the most mature drone could produce. The conclusion to be drawn was obvious: queens that return with the mating sign from a single mating flight could have been mated with a number of drones.

In my opinion this discovery is the most important in the sphere of apicultural science and breeding since Father Dzierzon first noted the fact of parthenogenesis in 1835. Parthenogenesis and the consequent genetic uniformity of the spermatozoa of a drone increases the dangers of inbreeding, but we now know that this is counteracted by Nature through multiple mating.

We are now faced with another question. When a queen on one mating flight, or more than one flight on the same day, mates with several drones, is there a mixing of the semen of the different drones? This in turn brings us to another point of the utmost importance in breeding. We now know that after mating the sperm do not mix together in a seminal fluid, but they tend rather to coagulate in clumps. We know too that the amount of sperm present is something like twelve times the capacity of the queen's spermatheca. It would seem from this that only those sperm pass into the spermatheca which are near the opening of the spermathecal duct. In other words, the spermatheca contains not sperm of all the drones with which the queen mates, but only of some of them as determined by chance. It is generally agreed that the queen mates on average with ten drones.

At the moment there is no clear answer to all this. Instrumental insemination is giving us a lead to the solution of the problem. From the strictly practical point of view, however, there is no real difficulty, for no serious breeder relies on matings which take place at random. He

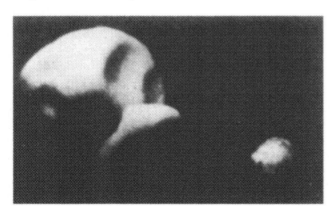

To the left, the actual quantity of semen extracted from a queen following a mating flight. To the right, at the bottom, a full spermatheca. Magn. ×17.

(After Ruttner)

has to rely either on a secure isolation apiary or else instrumental insemination. In an isolation apiary which is run for the purpose of queen breeding what is needed is a series of sister queens, daughters of a carefully selected breeder. The drones of these daughters manifest the hereditary factors of their grandparents, or more exactly, they show the characteristics of the colony headed by their grandmother. This means that there is a diversity among these drones, which is an advantage in that it gives us more room too for manoeuvre from the hereditary point of view. An absolutely rigid uniformity has no place in serious breeding. On the other hand, of course, it is possible by instrumental insemination to limit one's work to the drones of one definite colony. But there are always dangers latent here, since we never know in advance which line of queens will produce the best drones for breeding purposes. At one time Dr. U. Kramer thought that he had established this point with certainty and he formulated his breeding methods accordingly. Fortunately he was spared the worst results of his assumption by the complete unreliability of the mating stations at his disposal.

It should be mentioned here that by relying on random matings the progeny of a young queen can be partly 'pure' and partly crossed. Hence there is a possibility of developing a pure breeding line from a queen of this type, provided the selection is concentrated on the queens manifesting the greatest uniformity in their offspring. As is well known, the influence of a cross made some generations previously will manifest itself most clearly in the colour pattern of a queen.

Multiple mating of queens is without doubt one of the most important measures devised by Nature to preserve the vitality of the honeybee. At the same time it acts as a counter to the many undesirable consequences of parthenogenesis.

The Advantages and Disadvantages of Inbreeding

Inbreeding can rightly be regarded as the touchstone of selective breeding. It is the indispensable means of intensifying, fixing and standardising the desirable characteristics and at the same time weeding out the factors we do not want. Inbreeding is employed in all forms of breeding both in the animal and plant world. In fact in plants self-pollination is the most intensive form of inbreeding, far more so than other means of propagation employed by Nature. On the other hand, where the honeybee is concerned Nature uses every possible means of preventing inbreeding. This is clearly shown by the fact of multiple mating with drones at distances from the hive of up to ten kilometres.

As practical experience shows over and over again the most serious result of inbreeding is the progressive loss of stamina. This loss of stamina affects all the essential activities of the bees and even threatens the very existence of many colonies. The devastating losses about which we hear so much are due, largely if not entirely, to this loss of vitality. It is an insidious and illusive deficiency which always shows up mainly in unfavourable climatic conditions against which a weakened constitution has no resistance. Nature then takes control and weeds out the unfit. Lack of vitality shows up too in a reduced capacity for brood rearing, inability for self-defence and above all in an increased susceptibility to disease. While therefore a carefully controlled inbreeding is undoubtedly an essential requirement it must be used with great care in breeding the honeybee. Experience has shown that without this care even the most productive strains of bees can be ruined in a few generations.

Mendel's Laws of Heredity

This is not the place for a detailed account of the laws of inheritance formulated by Mendel. I will make reference only to those points which have a direct bearing on bee breeding. There are a number of excellent textbooks on the whole subject of heredity and genetics in all its aspects which any interested reader can easily consult.

Mendel's laws embrace every kind of living creature including the honeybee. However as I have already pointed out parthenogenesis and multiple mating, factors which we do not meet with in the general breeding of animals, play a decisive role in the genetics of the honeybee. Moreover instances from breeding of domestic animals which are quoted (not always from the best informed sources) are not valid in bee breeding, even when parthenogenesis and multiple mating are taken into account. In what follows I shall limit myself to the special case of breeding the honeybee and try to describe the problems, exceptions and difficulties we have to deal with. Mendel's own experiments in this sphere (he was a beekeeper) foundered because he had no control over the drones. Presumably he did not appreciate the influence of parthenogenesis and he had no inkling of multiple mating.

We know that Mendel was a very keen beekeeper, but as we have just mentioned, his attempts at cross-mating with his bees were doomed to failure from the outset. On the other hand his careful experiments with peas combined with his natural mathematical genius enabled him to establish the general laws of heredity. His findings clearly proved that the genes, or the factors as he called them, on which heredity depends were constants and not miscible like a fluid, as up to then had generally been held. The description 'mixed blood', which is still used occasionally today, has no justification in fact. In heredity there is no question either of a mixture or of blood, but of an exchange and a reciprocal influence of individual genes which determine the development of the characteristics of a living being. The laws of heredity are called Mendel's laws and the designation Mendelism is given to the principle governing hereditary characteristics because Mendel was the first to point out that individual genes, on which heredity depends, remain unaltered from generation to generation. This is a general rule of universal validity in spite of the limitations and many variations of which Nature alone is capable. We can see confirmation of this in daily experience. There are no two men, animals or plants which we could describe as completely identical, although they show all the traits of their species and usually reflect a family likeness.

The basic elements of Mendel's laws can best be grasped by citing the classic example given in all textbooks on breeding. The simplest case is that of a cross between the red and white garden flower, the 'Marvel of Peru', *Mirabilis jalapa*. The first results from this cross (the F1) are neither red nor white but rose coloured.

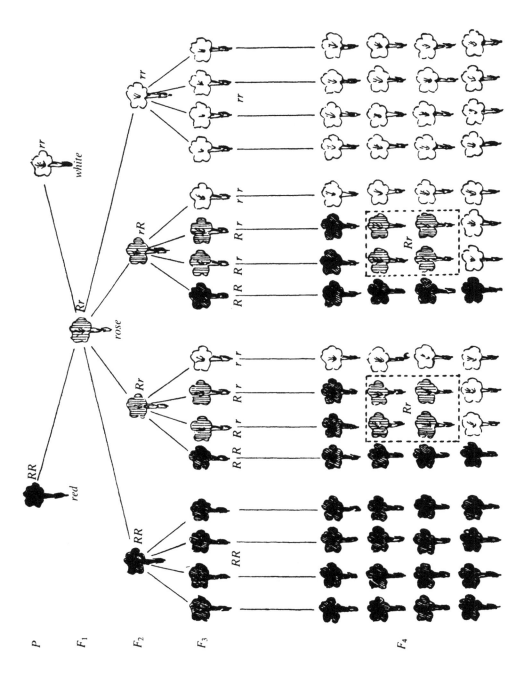

A cross between two varieties of Mirabilis jalapa, *demonstrating an intermediary inheritance.* (After Armbruster)

Mendel discovered that if rose is crossed with rose, in the second generation (the F2) segregation occurs. Of four individuals two are again rose; one is red, and one is white. These last when crossed among themselves, that is inbred, produce offspring like themselves, that is, pure red or pure white. On the other hand the rose coloured individuals when crossed inter se produce individuals in the same proportion as in the F2, namely two rose, one red and one white.

Here we have in a nutshell the basic principles of Mendelism: in the F2 segregation occurs in the proportion of 1:4, 2:4, 1:4 for every pair of characteristics. In the example just given the rose coloured plants are recognised at once as crosses, whereas the red and white individuals when inbred will always breed pure red or pure white. Mendel himself designated the hereditary factors with letters. The factor for red was given the capital letter 'R', that for white the small letter 'r'. All the germcells of a pure red plant always contain the factors 'RR', those of a pure white plant 'rr', but the cross contained either 'Rr' or 'rR'. In this cross the F1 is a sort of intermediary, being neither red not white but rose. This intermediary phenotype in the F1 is not an invariable occurrence. A further example will clarify this point.

A cross between a banded and a white garden snail produces a white snail in the F1. In this snail the factor for white prevails over the factor for banded and is known as the dominant. In this case when the segregation takes place in the F2, the result is three white individuals and one banded. This banded snail with the recessive factor and the formula 'rr' does not segregate again; it always breeds pure. Of the three white snails two are crosses which will segregate in the same proportions as in the F1. The one pure white, which will always breed pure, can be determined only by means of inbreeding. Hence we have here two outwardly recognisable colour patterns instead of the three in the case of the Marvel of Peru flowers. The proportions are 3:4 white and 1:4 banded, but the genes are still 1:4 'AA', 1:4 'aa', and 2:4 'Aa'.

So far we have considered Mendelism with only two genes in question. Now we come to a more complicated but very instructive example, a cross between two varieties of Antirrhinum, but with the dominant factors in only one pair. Both varieties differ in colour, one white and one red, and also in form, one normal and the other tubular. In the F1 an intermediate variety appears as in the case of the Marvel of Peru, namely a rose coloured flower but in which the normal form of the flower is dominant over the tubular form. As shown in the diagram by inbreeding from this F1 we obtain from the 16 descendants the following results in the F2: 3 red individuals with the normal form; 6 rose and normal; 1 red and tubular; 2 rose and tubular; 3 white and normal and 1 white and tubular. Among these descendants we find two new combinations, that is, red and normal form and also white with tubular form, both of which breed pure. Moreover the two original forms appear in their unaltered original purity. The pure white and the pure tubular individuals are immediately recognisable as are the double crosses with the rose colour. But only further inbreeding can tell us which ones with the red or white colour and normal form will breed pure.

It seems to me that this example brings out very clearly the kind of segregation which we must look for in the cross-breeding of the honeybee. And indeed practical

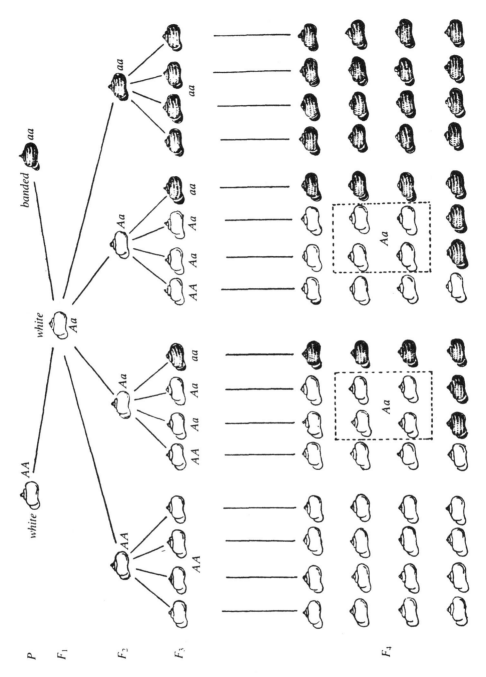

A cross between two varieties of snails, in this case an instance of dominance of white over banded markings.
(After Armbruster)

experience shows that this is not just mere theory. Uniformity of external characteristics in bee breeding is no guarantee of purity of inheritance. Only further matings inter se can produce the individuals which will breed pure. A minimal number of 16 is therefore absolutely essential to obtain the types indicated.

Many other examples could be cited of more complicated crosses and segregations in the F2, but I do not think there is any point in doing this when we are concerned here with the breeding of the honeybee. The three examples given above show clearly the essentials of Mendelism as well as their significance for practical scientific breeding. Mendel showed from the figures he obtained that the proportions of the individual factors remained constant and that they did not mix but kept their individuality from generation to generation. Moreover and this is of particular importance for the purpose of practical breeding, although the general rule holds good that these individual factors breed true and independent of each other, this does not happen in every case. Hence we are able to breed individuals with new combinations of hereditary factors. As we saw in the case of the two kinds of Antirrhinum it is possible to effect an interchange and combination of red or white colour with the normal or tubular form. Obviously this possibility is not restricted to external features only but extends to most hereditary dispositions.

All this shows that cross-breeding when properly conducted offers the possibility of synthesising at will the genes which are at hand and also of producing new combinations which better meet our needs today. In addition the more factors at our disposal in a cross, the more numerous are the segregations and the greater variety there is among the different individuals and their external forms.

In a four-way cross we obtain sixteen different kinds of germcells or gametes which are produced in the F1, and represent 256 possible ways of further fertilization. No less than 16 of these which segregate in the F2 are completely new types or combinations which breed pure. The possibilities of such a four-way cross can be illustrated by the classic example of a cross between two types of barley. The two types differ in four characteristics:

1. upright and drooping   (habit)
2. hooded and bearded   (awns)
3. two-rowed and four-rowed (ear)
4. white husk and black husk (colour)

In the diagram on page 29 in the middle is the F1. It is 1) drooping, 2) hooded, 3) two-rowed, 4) black husk. In the other groups there are the 16 homozygotes, of which 2 represent the original pair and the other 14, new combinations.

Inbreeding produces a regrouping of the characteristics in the F2 which provides us with the key for creating new combinations of characters. It is in this way that over thousands of years the numerous valuable varieties which we now possess have arisen. These have come about either by planning or by chance in every form of life, animals, plants, the various berries and fruits. So too in the distant past, chance crosses of the varieties of wild grains produced the types we now have, the originals of which are no longer extant. It is probable that the development of these types took place over a period of millions of years and not just at one particular moment. Further developments on these lines are open to us today through careful selective breeding.

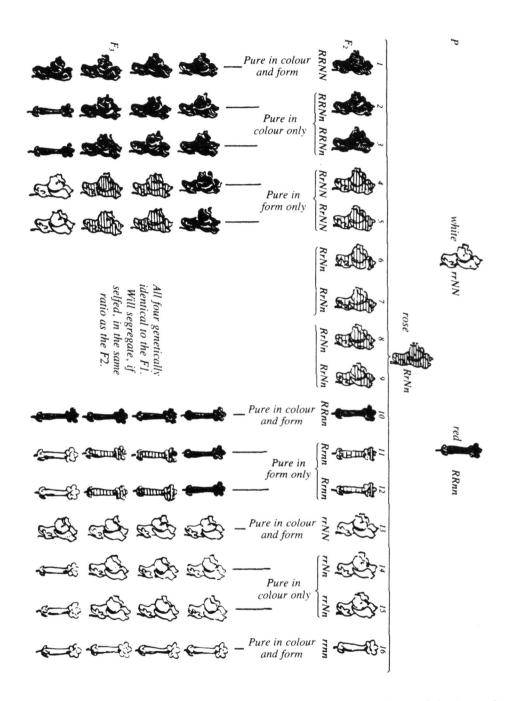

Antirrhinum — *An interesting case of an intermediary inheritance in colour and dominance in form on the one hand and recessiveness on the other.* (After Armbruster)

The most remarkable achievements have up to now been among plants. Here the breeder has at his disposal not only an almost unlimited number of specimens for possible selections but also a rapid following of one generation upon another. These two advantages do not hold for the domestic animal world at large. However as we shall see the honeybee is an exception (happily to our benefit), although not to the same extent as in plants.

An easy way for the wouldbe breeder to have a bird's eye view of all the crosses possible in an F2 is to construct a sort of draught board. On the top edge are printed the male gametes or their symbols and on

The classic case of a barley cross, resulting in 14 distinct new combinations.
(After Armbruster)

the left edge the female gametes or their symbols. In the squares can be written the symbols or letters, first those taken from the top edge and then those from the left hand edge of the board. In this way none of the possible combinations can be overlooked. In the case of the flower Marvel of Peru there are four possible combinations: 'RR', 'rr', 'Rr' and 'rR'. The last two produce different offspring although they are the same genotypes.

When we come to a cross involving two pairs of characteristics we need 16 squares; with three characteristics 64 squares; with four 256 squares, as can be seen from the diagram on page 34. On the diagonal drawn from the upper left hand side to the lower right hand side we have the individuals which are doubly pure, on the other diagonal those which are double crossed. These require only selfing to make the segregation exactly as in the F1.

Racial Purity in the Light of these Laws

One constantly hears people talk about a 'pure' Carnica, a 'pure' Ligurian and so on. But in the light of Mendel's laws it is clear that this applies to a relatively few characteristics, such as the red and white colour of the Marvel of Peru. In reality there is scarcely any individual which can be described simply as a pure race, apart from some of the simpler forms of life which multiply by simple binary division. In such creatures the complete set of genes is handed down untouched, as it is where self-propagation is the normal method of reproduction. In addition a favourable or unfavourable environment can hinder or favour development and growth in these forms within the limits imposed by inheritance. Hence there is no question of selection with these simple forms of life. It is only where we have hereditary complex forms of life that it is possible for us to make use of selection and careful mating to intensify and fix the characteristics we require and also weed out those which are undesirable. It follows from all this that the only possible way to make progress in the plant and animal world is by crossbreeding. This entails a clearly defined aim for our breeding, a corresponding selection and then line-breeding. The history of the development of living forms and practical experience confirms this conclusion. Yet, as we have already remarked, a high degree of racial purity nearly always entails a fundamental loss of vitality.

A point to be borne in mind in connection with this is that because of multiple mating in the case of the honeybee we have queens which are at one and the same time pure mated and mixed mated. The progeny of such mixed mated queens is partly pure and partly crossed. Hence by careful selection we can breed queens of absolutely pure type from differently coloured descendants. This conclusion is derived from theory but is confirmed by actual results.

The Application of Mendelism to the Honeybee

We have sketched briefly the laws of genetics derived from Mendel's experiments. These form the fundamental principles of our work in the breeding of the honeybee. I do not think there would be any advantage in dealing here with some of the more complex examples of inheritance, problems which do not arise with the honeybee. Once more I must refer the interested reader to the many textbooks on the subject. I would like to concentrate now on the special place held by the honeybee in the context of Mendel's laws.

In spite of parthenogenesis and multiple mating Mendel's laws apply in full to the honeybee. The only exception is that the results of segregation and numerical ratios are different. But the hereditary factors are preserved in their original purity exactly as in other forms of life, not however in as straightforward a manner as in the animal and plant world but in a more roundabout way.

We do indeed obtain an F1 which partly corresponds to Mendel's laws, but this progeny is not descended from one father as in other life forms but from an unknown and undetermined number. As a result the female descendants of each queen are only half-sisters not full sisters as is normally the case. Each individual bee has of course only one father. At the same time each colony is composed of a series of groups of genetically identical super sisters, the daughters of one particular drone which mated with the queen. A drone has no sons, only grandsons; he has no father and his hereditary factors correspond to those of his grandparents. Hence in a first cross we have no F1 drones, they appear only in the subsequent generation.

The two factors of parthenogenesis and multiple mating, especially the former, prevent our obtaining in accordance with Mendel's laws, the segregation and production of new forms in the F2. True, we have a segregation in the female progeny of a first cross, but of the male progeny only in the F2, and even then it is among an indeterminate group of half-brothers. We can therefore set out a draught board with the different gametes of the queen on the left hand border and a similar one for the drones on the upper horizontal line. In this way we can show the respective gametes of the sons of an F2. Both series of gametes, those of the mother and those of the father, are identical in their makeup. Thus in the case of the honeybee the symbols on the upper line represent male individuals, each able to transmit ten to eleven million spermatozoa of identical hereditary composition corresponding to that from which the drone itself sprang. This is one of the unavoidable consequences of parthenogenesis.

It should of course be clearly understood that the matings about which we are speaking are no mere random matings with drones of unknown origin. No serious breeder could rely on such chance happenings. The crosses we are concerned with are effected either by instrumental insemination or by selected matings in a completely isolated mating apiary. In addition we need to have at our disposal only races and individuals with clearly distinctive features and physiological characteristics. Breeding on modern scientific lines cannot be pursued without these basic requirements.

In order to make clear the fundamental principles demanded by breeding along the lines laid down by Mendel, I would like to refer to the table on page 34 in which there is an example of a three-way cross brought about by a cross between two types in which the colours are completely contrary, that is the Nigra and the Aurea. Although it is a hypothetical case it does indicate the essential guidelines we have to observe in the breeding of the honeybee. The purpose of this cross is to transfer the wingspan and the wingpower of the Nigra to the Aurea, obviously something of great advantage to us. Other characteristics come into consideration but they can be fixed without much difficulty.

Following the custom started by Mendel we identify the characteristics in question by using letters of the alphabet. The dominant characteristics are given capital letters and the recessive small letters. So in the case under discussion the Nigra is identified by 'SS' for black, 'LL' for long wingspan, and 'dd' for sparse hair. The Aurea is designated as 'ss' for golden colour, 'll' for short wings, and 'DD' for dense hair. The pure Nigra therefore has the hereditary characteristics in the form 'SSLLdd', the Aurea 'ssllDD', and the F1 'SsLlDd'. As experience shows, the characteristics of the Nigra are dominant over those of the Aurea as is the case with nearly all the other races. We are not concerned here with the various intermediate forms.

As can be seen from the table the segregation of the three pairs of characteristics produces eight different kinds of gametes, which in turn produce by the appropriate selfing 64 different fertilization possibilities. But of these we are concerned only with those on the diagonal from top left to bottom right, as only these give us the six new combinations and the two original parent races with their original identical characteristics. The other six individuals are also racially pure, of course only as regards the characteristics in question. But they do incorporate a variety of linked characteristics which breed pure. Among them is the type 'ssLLDD', that is the Aurea with long wingspan.

To obtain these new combinations we need a minimum of 64 young queens. In view of multiple mating only rarely will a pure mating result with 'SLD' drones from the eight groups of drones required. This occurs rarely even in the most secure isolation apiary or even if we were to handpick the desired number of drones identical in their external characteristics and bring them for the mating. What does save us all this trouble and work is instrumental insemination. From a number of young queens we can likewise pick out a few with 'SLD' characteristics which we can identify by their external form.

Although as I have said this is a hypothetical case, it is by no means purely imaginative but is one based on actual experience. It must be emphasised however that when we are dealing with a multiplicity of characteristics (which is what we meet with in practice), in the manifold crosses there occurs an almost unpredictable number of segregations and breeding possibilities. True, the number can be worked out mathematically, but in reality it is almost limitless. However in breeding the honeybee we have one great

NIGRA   AUREA

P = SSLLdd   ssllDD
F_1 = SsLldD

S = black   s = golden
L = long wing   l = short wing
d = sparse hair   D = dense hair

F_1 ♀ / F_2 ♂

	SLD	SLd	SlD	Sld	sLD	sLd	slD	sld
SLD	SSLLDD							SsLlDd
SLd		**SSLLdd**					SsLldD	
SlD			SSllDD			SslLDd		
Sld				SSlldd	SslLdD			
sLD				sSLlDd	ssLLDD			
sLd			sSLldD			ssLLdd		
slD		sSlLDd					**ssllDD**	
sld	sSlLdD							sslldd

On the diagonal line from top left to the bottom right are situated the new true breeding combinations. The two original parents are indicated in bold type. On the other diagonal are situated the multiple heterozygotes.

advantage, one which is shared in greater degree by the plant breeder, that is, we have at our disposal a very large number of individuals from which to make the selection and also a quick succession of generations. These are two important advantages denied to the animal breeder. Yet in spite of these benefits we can hardly ever obtain the ideal combinations. But as experience proves we can get approximations to the ideal which give us worthwhile results. These approximations lead step by step to others, so that gradually we do approach the ideal for which we are aiming.

Once we have established a clearly defined goal at which to aim and the application of a measure of inbreeding corresponding to that goal, cross-breeding in the honeybee puts unlimited possibilities at our disposal.

The dominant role played by parthenogenesis might seem to indicate that Mendel's laws are not applicable to bee breeding. This is a mistake. The origin of drones from a virgin mother and multiple mating are only apparent obstacles. In fact it is these two factors which enable us to arrive at a synthesis of new genetic characteristics with less difftculty than in other forms of breeding. Males with their appropriate genetic makeup are always at our disposal in greater numbers. Moreover, thanks to parthenogenesis, the millions of sperm produced by an individual drone are all identical from the genetic point of view. This identity provides the maximum stability of hereditary factors in the offspring. Because of the haploid transmission a drone can transmit only those characteristics which were present in the egg from which it originated. With the drone there is no dominance, nothing concealed behind the external appearance of purity or the mixed colouring of a cross. It can never wrap itself up in transitional colours; hence it is possible by instrumental insemination to select drones which have the genetic form of 'SLD', that is individuals which have uniform yellow bands, maximum wing length and thick overhair. Or, if normal mating is desired one can select the drones by hand to give the required number for the mating station.

Without these theoretical and practical possibilities the application of Mendel's laws to bees would be denied us. Nor would we be able to pursue the breeding and fixing of new combinations, and thus progress would be impossible. We would therefore have to resign ourselves for ever to the present conglomeration of races of the honeybee.

Chromosomes and Reduction Division

Mendel knew nothing about chromosomes and cell division. But as his writings show he had an inkling that the units which are the carriers of the characteristics and are joined together as a result of a cross, separate in some way and are handed down independently. The advances made in microscopy and cytology since Mendel's time have confirmed his hypothesis. Although the process of heredity is common knowledge today, it will be helpful to set down the main points about chromosomes and cell division so as to get a better understanding of what occurs.

It was about 1838 that it was definitely established that the unit of structure and function of all living beings, animals and plants alike, was the cell. This discovery must be regarded as one of the most important in the history of biology. The structure of every cell is fundamentally the same. It consists of a living substance, the cytoplasm, in which lies the vesicle-like nucleus. The bodies of living beings are built up of cells which in spite of their variety of functions are all basically the same. Sense organs are composed of sense cells, nerves from nerve cells and so on. However much they differ in form, appearance and function they all have two essentially identical factors, they consist of a cytoplasm and a nucleus. If we trace back the different kinds of cell possessed by a fully developed living being to the egg from which it arose, the further back we go, the more we find that the cells are similar to one another. It is obvious that the original source of development, that is the egg cell, contains in itself all the future possibilities and characteristics of any individual living being consisting as it does of millions of cells obtained by constant division. Under a microscope with x 1,000 magnification we are able to see the chromosomes which contain the actual genes. In the last analysis the chromosomes are the organs of hereditary transmission and cell division the means by which a living being develops.

Every kind of living being has a fixed number of chromosomes. Among honeybees the queen and the workers have 32, the drone has 16. *Apis indica* or *cerana* has the same number as the various forms of *Apis mellifera*. But the giant bee, *Apis dorsata,* like the dwarf bee, *Apis florea,* has only 16 chromosomes in the female and 8 in the male. Chromosomes usually vary in size and shape, but are invariably alike in the case of an individual species and can only be identified at a particular stage of cell-division. Moreover the paternal and maternal chromosomes will pair only with chromosomes of identical size and shape. This is obviously so, to ensure the dual development of the particular characteristics. Technically these reciprocal genes are termed allelomorphs. Under a microscope chromosomes are usually seen in a tangled conglomeration. However at the reduction division they are marshalled in reciprocal pairs. The segregation of the chromosomes and the reduction from a dual to a single set takes place in the egg and sperm shortly before fertilization. This reduction of the

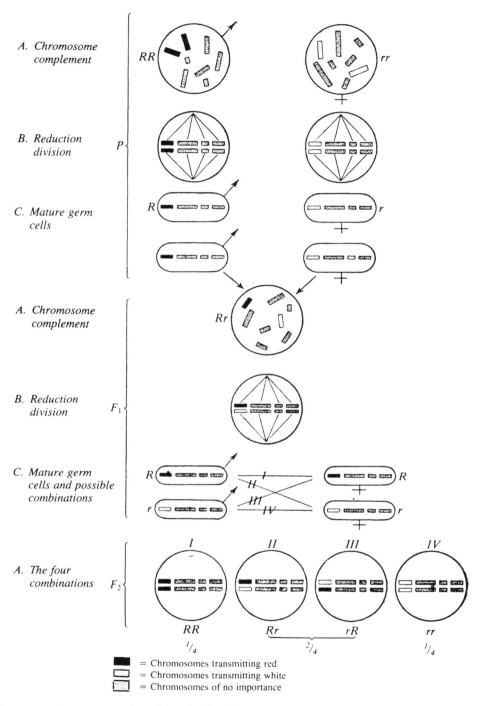

Diagrammatic representation of a reduction division (After Armbruster)

chromosomes to half their number is an essential provision for otherwise their number would double with each generation. The process of a reduction division is graphically indicated in the diagram on page 37. Every characteristic in living things, sexually reproduced, is determined by two chromosomes, one derived from the male and the other from the female. The drone sets an exception. At the fertilization the chromosomes unite in pairs but are again separated at the following reduction division and subsequently reunited in a series of new combinations. As already indicated, Mendel postulated a modus vivendi of this kind long before the existence of chromosomes, the process of their segregation and re-combination was recognized — an instance where the mind and intellect pre-supposed what was subsequently by means of modern optics in actuality ascertained and indubitably established.

The Reciprocal Influence of the Genes

In the foregoing pages we have discussed the relationship which exists between the genes in the chromosomes and the definite characteristics which depend on them. The example of the flower, the Marvel of Peru, might give the impression that every gene produces one definite characteristic without the influence of other genes, or to put it in another way, that every gene, for example that for the colour red, has only one exclusive function which it performs in complete independence of any other gene. This is very far from the case, as individual genes affect a number of characteristics in most cases simultaneously. This must be regarded as the norm, and any case of one gene having an influence on one single characteristic as the exception. In fact we have to concentrate more on the interplay of the genes and their combined efforts in the production of each characteristic, although in the last analysis it is to one definite gene that we have to assign the end result.

A classic example of the team work of genes is afforded by the transmission of different forms of comb in poultry. This example shows up the surprises we meet with at times in breeding. The four different kinds of comb are shown in the diagram on page 39. A cross between a pea-shaped comb and a rose-shaped comb produces a walnut-shaped comb in the F1. When we cross them inter se we obtain in the F2 9 individuals with walnut, 3 with rose, 3 with pea and 1 with a simple comb. Here then we have the genes for pea combs and rose combs producing two further forms of comb, walnut and simple. These last forms are found in other races but not in the two parents with which we are dealing. Although we cannot expect to see such comparable results in bee breeding, surprising things do happen, not only as regards colour and other external characteristics but also physiological ones as well.

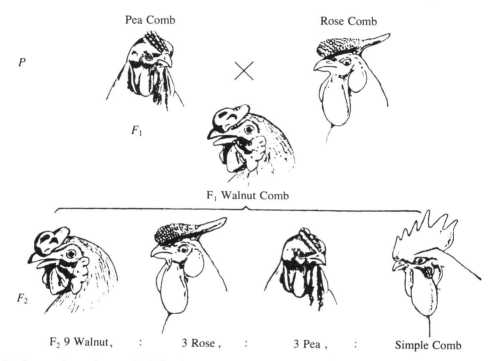

Comb types in poultry — depicting how a cross can give rise to characteristics not manifested in the original stock. (After Goldschmidt)

It is well known that in the case of mice and rabbits a few genes produce almost endless shades of colour of the fur — grey, black, blue, yellow and silver. In the case of the honeybee the colour varies from jet-black, black, brown, lemon, orange-yellow and a deep aurea, while the hair can be white, grey, yellow, brown and jet-black. With mice we find also a colour-change gene, a colour-fixing gene as well as genes one of which produce a glossy and the other a mat finish. With bees we have only differences between black, brown, yellow and aurea. But within that colour range we have the jet black of the Nigra, the leather colour of the genuine Ligurian and the orange of the Middle East races which is distinct from the aurea of the Italian. Moreover we cannot overlook the many shades of intermediate colours which occur. Although these are due to fewer basic colours than is the case with mice and rabbits, yet there is no doubt that they are caused by a reciprocal influence of a series of hereditary factors or genes.

The suggestion has often been advanced that the extreme yellow colour, usually called aurea, is due to a cross between the Ligurian and the Cyprian. Our breeding experience has shown however that this deep golden colour, which normally covers four or five segments, is always the result of a cross between the dark west European bee and the Ligurian. At the other extreme we have the jet-black Nigra. This seems to be a case of atavism, a throwback to the primordial ancestor of the west European group of races, the Intermissa. The Nigra and the Intermissa are not only identical in colour but in almost all other characteristics.

Results secured in the crosses indicated

	a	b	c	d	e
In a single cross	1	$2^1 = 2$	$(2^1)^2 = 4$	$3^1 = 3$	$2^1 = 2$
In a twofold cross	2	$2^2 = 4$	$(2^2)^2 = 16$	$3^2 = 9$	$2^2 = 4$
In a threefold cross	3	$2^3 = 8$	$(2^3)^2 = 64$	$3^3 = 27$	$2^3 = 8$
In a fourfold cross	4	$2^4 = 16$	$(2^4)^2 = 256$	$3^4 = 81$	$2^4 = 16$

a) Number of pairs of characteristics
b) Number of distinctive gamets formed by a F_1
c) Number of possible combinations
d) Number of genotypes
e) Number of externally distinctive genotypes — corresponding to the phenotypes in a F_2, if each characteristic proves fully dominant.

Polymeral Transmission

We must now bring in another important side of the interplay of the genes. When a particular characteristic makes its appearance as the result of a greater or lesser number of genes combining together, we have what is known as polymeral transmission. This refers especially to characteristics of a quantitative nature, such as length, weight and size, which in this way are accentuated and brought to light. They are of great value in general breeding and play an important part in our bee breeding work. We need, for instance, a bee which is strong on the wing and one which has a long tongue reach where red clover is at hand. But in both these cases it is a matter of abilities which are more of a qualitative nature. The fixing of the colour is without doubt due to the interaction of a combination of genes, and in general all our efforts are aimed at improving the quality of the characteristics. The performance of the bee is with few exceptions due to special qualitative factors; good temper is a classic example of polymeral intensification.

It is well known that between extreme good temper and extreme aggressiveness there are every imaginable variations. The same applies to resistance and proneness to disease. It has been clearly demonstrated that there is hardly a characteristic of the honeybee which is not due to a combination of genes. The line-breeder, willy nilly, bases his whole work on this fact. The intensification of any given characteristic is possible only by means of a summation of a series of allelomorphic genes.

The respective genes are not only in pairs but normally in a whole series. Each one of the pair of genes produces a different combination and as a rule these different combinations are the steps in the development of the same characteristic. If for example, 'A' brings about a definite characteristic, then we have A_2, A_3, A_4, etc. of which each allele represents a further step in the process of intensification. Wing length and tongue reach, cubital index, and many other characteristics are the clear result of quantitative alleles. Although we have to recognise the important part played by these quantitative alleles in breeding, there is no doubt that the qualitative alleles play a far more important role. There are very considerable differences of size in the different races of bees which are due to heredity. Variations in size within the same race are usually due to favourable or unfavourable conditions in the environment. This is very noticeable in the rearing of queens.

Linked Characteristics

Another fact we have to reckon with is that there are characteristics which are inseparably linked with one another. Every living creature possesses a definite number of chromosomes in accordance with its species. The fruitfly *(Drosophila melanogaster)* has only four pairs of chromosomes and yet more than 500 genes have been discovered. Queen bees and workers have 32 chromosomes, while the drone as a haploid individual has only half that number, viz. 16. The number of genes is unknown but it must run to several thousands. Hence we find in the individual chromosomes a series of genes which not only exercise their influence on a broad spectrum of characteristics, but are also coupled together in such a way as to be completely, or almost completely, inseparable. To get over this impasse Nature has provided an escape route in the form of an exchange of adjoining chromosomes called a 'cross-over'. It is an interesting subject but from our point of view is more of academic value. What we have to bear in mind is that some genes are inseparably linked and that a cross-over does take place though seemingly seldom.

We must mention one further point. Appearances occasionally seem to indicate that two or several characters are inseparably linked. This happens when a certain combination of characters occurs more frequently than would be predicted by the laws of chance, although a segregation of individual characters does happen at times. In such instances the ratios secured will provide the correct explanation.

The Limitations Imposed by Heredity

Speaking generally we have to take into account the enormous influence exercised by environment in the development of genes, in the plant and animal world. Yet environment has no influence directly on the genes themselves but only on their development. It is not necessary to cite examples as everyone is acquainted with the fact from his own experience and observation. A plant or animal in the most congenial environment will develop into a first-class specimen, but placed in the wrong environment will become an individual wanting in various ways. But in both cases the the development takes place within the limits imposed by the genes there present.

To a certain extent we have to take note of this problem of the restrictions imposed by the actual genes. The life cycle of a colony, especially during the breeding and development from egg to fully-fledged bee, takes place in a temperature which remains almost constant and in a humidity which can be regulated by the colony according to need. As long as there are sufficient stores of honey and pollen available, the seasonal honeyflow plays little part in the development of the individual bee. If stores begin to run low, a normal colony will take counter measures in the form of restrictions so as to avoid a state of complete dearth of stores. These restrictions are the interruption of brood rearing and the expulsion of the drones. A colony has the ability to protect itself against the results of unfavourable circumstances. It is equipped with wonderful instinctive powers to overcome well-nigh impossible obstacles it occasionally meets with, and now and again caused by the incompetence of the beekeeper. However there are times when these powers are not up to dealing with the situation. When these situations do occur, then the hereditary powers of endurance of the bees and the fertility of the queen are adversely affected. Likewise the virility of the drones is lessened. We have here instances of wide genetic fluctuations in the development of certain hereditary dispositions, particularly in the reproductive members of a colony. Fluctuations of this kind are often difficult to assess.

Mutations

Sudden changes in the genetic makeup of a living organism are called mutations. Darwin based his theory of evolution to a large extent on changes of this kind. Armbruster and other scientists were at one time of the opinion that we could make use of mutations to open up other breeding possibilities. It was then customary to distinguish between mutations which were advantageous and those which betokened a loss. Today however we know that all mutations are detrimental, that they bring with them an impoverishment which shows itself mainly in a loss of vitality and occasionally in anatomical abnormalities. Mutations which are dominant are more often than not lethal. For example, in the Irish breed of cattle called Dexter we have a case where a mutation produces a 25% ratio of stillborn calves. In the case of mice, yellow colour is due to a lethal factor. Of individuals which possess this lethal factor only those survive which are of mixed descent. The pure yellow mouse dies in the first stages of its existence in its mother's womb. However, the greatest number of mutations known in the case of the honeybee are mainly of harmless nature. Their effect is limited to a partial restriction or blocking of the effects of a particular gene, as for instance, we find in the colour of the eyes in bees. The effects of mutations are clearly brought out by the list of 30 mutations so far identified, of which 18 affect the colour of the eyes, 5 the wing-formation, 3 the colour of the dorsal segments, 2 the colour of the hair, and 2 behaviour. The variation in the colour of the eyes is easy to explain. It is obviously due to a progressive loss of the different alleles which determine colour of the eyes. White eyes, which do often appear, are due to a loss of all the colour alleles. The colour 'Cordovan', a leather colour, is widely regarded as a mutation. In my opinion it is rather a matter of a basic colour, a dominant one, which asserts itself in the heterozygotic state. On the other hand, sterile eggs are certainly due to a mutation; this seems to occur only in the west European races. A mutation always means a loss of something; a reversal process to the previous condition has never been achieved — in fact it is not possible.

A mutation which means the loss of a gene can never be regarded as a gain. It could possibly be seen as a gain if the loss of a gene brought about the development of a situation in which a recessive factor was an advantage. Such a case occurs in fruit trees. Now and again a tree produces a branch of which the fruit, as a result of a mutation, does not completely correspond to the normal characteristics of the particular variety. If desirable, the mutation in such an instance would be a benefit, for the new type could be propagated by grafting. However mutations of even this sort always bring with them a lessening of stamina, which is the general rule with mutations.

In the case of bees experience shows that in addition to those already mentioned there are a number of mutations of a lethal nature which cause problems in practical beekeeping. Susceptibility to one or other of the diseases which affect the brood or adult bees is doubtless largely due to a change in the genetic makeup.

Lack of vitality lies at the root of nearly all our problems of beekeeping. Inbreeding leads to an intensification of small mutations and unavoidably to loss of vigour. Large mutations and marked deformities do appear in bees, but these are naturally excluded from propagating farther.

As we have already remarked mutations are changes in the genes which appear suddenly and are permanent. We can produce them artificially by means of ray-treatment, chemicals and extremes of temperature, but these are chance occurrences and can seemingly never be predicted. In the development of every living organism there is a crucial stage in which it is most exposed to dangers. In the narrow world of the bee, made secure by Nature for her breeding and development, the bee is almost proof against mutations. It is likely however that the real Achilles heel of the bee is to be found in the spermatozoa held in the spermatheca of the queen before they are united with the eggs. Take the case of the hermaphrodite; among bees we occasionally come across these individuals with both male and female characteristics. This seems to be the result of the failure of a gene in the sperm which prevents the full development of the female organs. This is clearly the most obvious explanation, as the egg would have developed into a drone without fertilization. Although other explanations have been suggested none of them seem convincing. Hermaphrodites have been produced by cooling the eggs, or more precisely the sperm. Experiments have shown that by lowering the temperature of a queen, the sperm in her spermatheca are killed, yet without damage to her ovaries or her laying ability. The result is a drone-laying queen.

Mutations are the hobby-horse of more than one geneticist concerned with bees; they have been assigned scientific formulae in the hope that a study of them would throw light on some aspects of heredity in the bee. But research on mutations entails much work and is only feasible by means of instrumental insemination. Moreover such creatures are destined by Nature for early extinction, and it does not seem that we can expect much gain from them for practical purposes.

Synthesising New Combinations

From what has been said it can be seen at once that mutations do not lead to any progress but always to an impoverishment of the hereditary makeup and a lessening of the power of adaptation to the conditions of life in the individuals concerned. Anything which cannot survive the hazards of life is soon weeded out by Nature. Nature has no time for the halt and the weakling. It is always the creature of the wild, be it animal or plant, which is the strongest of its kind, and from the point of view of practical economics not necessarily the most productive. As Darwin pointed out, 'Survival of the Fittest' is Nature's unrelenting law, and she admits of no weakening of this principle. It is thanks to this ruthless attitude of Nature that we have an invaluable fund of geographical races of bees at our disposal for the untold possibilities of breeding.

As Mendel possibly suspected we today are in the position of being able by means of cross-breeding to produce almost at will a synthesis of hereditary factors of the utmost economic value. This is the only way in fact to make any real progress in the breeding of the honeybee both from the theoretical and practical points of view.

The breeding of domestic animals and of plants has already produced countless examples of successful achievements of this kind.

The plant breeder is in a much more advantageous position than the animal breeder in this connection. He can produce without any trouble an unlimited number of individuals in the F2, and so has an almost unlimited choice for bringing about ideal combinations. As I have already pointed out we bee breeders find ourselves in the same happy position. True, we do not have all the advantages of the plant breeder, but there are possibilities open to us which are closed to him. The results already secured in the synthesising of new combinations clearly demonstrate the possibilities that await us in bee breeding.

We are concerned here not only with possible improvements in the performance of the bees but also with their greater resistance to disease, since without such a corresponding resistance maximum performances are not attainable. I must emphasise that I am not speaking about immunity to disease but of a high degree of resistance. Allow me to quote an example from plant breeding: a virus infection of the sugar cane in Java brought about enormous losses. By crossing the susceptible cultivar with a wild variety, a new combination of genes was effected with the result that not only did the new strain show almost an immunity to the disease but also increased production fourfold. Our own experience in the fight against Acarine convinces us that we are not considering Utopian flights of fancy or indulging in mere theorising or wishful thinking.

The Limits Imposed on Us

We mentioned previously that there are 'immeasurable' possibilities for breeding put at our disposal by the great variety we find among the races of bees. But there are also limits beyond which we cannot go and which we must always bear in mind. We can indeed use all the vast possibilities of combining genes, but we cannot produce something which in some form or other was not already there. There is no such thing as 'spontaneous generation' of new genes nor a spontaneous progressive accumulation of valuable genetic dispositions, but an ever present possibility of a genetic erosion.

As we have seen mutations betoken either a partial or a total inactivation of a gene. In the same way the intensive pure breeding and the massive propagation of a very limited number of strains — as is much practised today — can mean only a noticeable impoverishment of the total gene supply of the honeybee, and this decline is obviously not confined to just the undesirable characteristics. The whole subject of these modern trends is one which demands our most careful attention. Plant breeders are today bemoaning the loss of wild varieties of corn possessing so many factors of economic value for breeding which were at their disposal thirty years ago but today cannot be found.

Sex Determination

Finally we must turn to the question of sex determination in the honeybee and the role played by the sex alleles and their bearing on breeding.

The characteristics and external appearance of all living beings is fixed by an interplay of pairs of genes. If the genes bring about the same characteristics they are called alleles. Thus the determination of sex in most animals is due to the two sex chromosomes known to geneticists as the X and the Y chromosomes. The female has two X chromosomes but the male has an X and a Y, that is XX for the female and XY for the male. Hence the egg cell can transmit only the factor X, while the sperm can transmit half with factor X and half with factor Y. It is probable then there will be an equal number of the sexes because an egg cell will be united with a sperm having either an X or Y factor.

However the sex chromosomes are differently assigned in the various types of animals. In the case of mammals the female has a uniform allele, that is XX. With birds it is the opposite, as it is with the Hymenoptera to which the bee belongs. But it is even more complicated with the honeybee.

As with all living beings there are in bees two factors which determine sex, but there are not only two forms, but a series of variations known as sex alleles. When two different sex alleles, an X and a Y, are united then we have females. But since the drone is 'virgin-born', in the sense that he has no father, he normally possesses only one sex allele. Intensive inbreeding leads infallibly to a restriction of the sex alleles and at the same time to an intensification of identical sex alleles. Such eggs are however viable and would develop into diploid drones. But this kind of larvae are discarded by the nurse bees shortly after emerging and so can never develop. How the nurse bees are able to distinguish these special larvae from the normal worker brood is a disputed point.

The following table shows the process of sex determination. The symbols have been chosen without any significance:

		queen		drones		descendants
sex allele	—	a/b	×	c or d	=	a/c-a/d-b/c-b/d
sex allele	—	e/f	×	g or h	=	e/g-e/h-f/g-f/h

All 8 descendants have different sex alleles from which females will develop. On the other hand the next diagram illustrates where there are identical sex alleles in 25% of descendants in the first instance and 50% in the second instance.

		queen		drones		descendants
sex allele	—	a/b	×	a or c	=	a/a-a/c-b/a-b/c
sex allele	—	b/c	×	b or c	=	b/b-b/c-c/b-c/c

This is of course a purely hypothetical example because in reality a queen does not mate with one single drone but always with an indeterminate number. In the case of random matings where drones of mixed racial origin fly, this rarely or seldom happens, certainly never with a brood loss of 50%.

Multiple mating is Nature's way of bringing about the required balance. But where it is deemed necessary to apply intensive inbreeding with the corresponding control of mating, high loss of brood and all the other drawbacks are unavoidable. In an ordinary commercial beekeeping enterprise such measures are of course not taken except when they are regarded as essential for obtaining special individuals for the purpose of further breeding work. In such cases we have to accept the fact that loss of brood and the other disadvantages will follow.

As has been noted when mating occurs at a mating station, any loss of brood caused by the presence of identical sex alleles is hardly worth consideration. The only cases where it is likely to happen is where intensive inbreeding is undertaken for scientific purposes, or when a sort of idealism leads to experiments undertaken in complete disregard of the disastrous results which follow. It was Mackinson and Roberts in the United States who in 1945 during a series of experiments in inbreeding with the help of instrumental insemination first met with such results. They both thought that the abnormal loss of brood was due to a lethal factor. Later, J. Woyke of Poland was able to refute this. He showed that if larvae which would have been removed by the bees were taken immediately after emerging and fed for two days on royal jelly and then given back to the nurse bees, such larvae developed normally. But the bees which emerged were drones of the diploid variety. Investigations showed that their reproductive glands were smaller and the amount of semen they possessed was only one eighth of the amount in normal haploid drones. In addition these diploid drones produced diploid sperm. This is yet another indication that the honeybee is a special case and a confirmation that we can use inbreeding only exceptionally.

A Word of Warning about Mere Theory

To set the foregoing considerations in their proper perspective, it is important to remember that the results achieved in the past in the breeding of domestic animals and of plants was largely obtained without the scientific knowledge of genetics we now possess. The many different breeders of domestic animals who made England world-famous some 150 years ago clearly knew nothing of the laws of cell division, chromosomes, cytology and the rest. The same must be said of the American genius Luther Burbank who 70 years ago had such amazing successes in plant breeding. Such persons as these relied on a sort of intuitive power, a kind of sixth sense. Today we have a wealth of knowledge at our disposal which these people did not have. Yet in spite of all this, in the field of breeding a certain measure of intuition is still a requisite.

The study of heredity embraces all living beings, indeed we are dealing with life itself. We know some of the rules which regulate the transmission of characteristics from one generation to another, but the multiplicity and diversity of the process remains a mystery beyond man's grasp. We have always to be content with but a partial analysis and a partial knowledge of the different aspects of heredity.

I have limited myself in this theoretical part to the main points which are essential to the discussion about breeding the honeybee. The literature on breeding is from the scientific point of view quite bewildering in its quantity, but from the practical side it contains little of real value. For us as bee breeders it can more often than not confuse rather than enlighten. In bee breeding we have to deal with the problems of exceptions to the general norms of bisexual heredity. The community character of bees, parthenogenesis, multiple mating and the fact that the individuals which carry the most important characteristics are not capable of reproduction — all these features prevent us from making too close comparisons between the animal and plant world and that of the honeybee.

Part II
Practical Possibilities in Breeding

Preliminary Remarks

Before proceeding further I must here put on record how much I am indebted to the work of Professor L. Armbruster on the theoretical aspect of breeding. He devoted almost his whole life to the study of the basic principles of bee breeding, taking as his starting point Mendel's laws of genetics. However a great part of his published work was done before multiple mating was established as a fact. Some of his findings were accordingly based on suppositions which were later shown to be erroneous. Likewise, some of his theories about the inbreeding of the honeybee do not accord with the facts. Whereas all the views which I have expressed in the previous section of this book are based on the knowledge and experience derived from over 60 years of practical bee breeding. Thus, for example, I was aware even before 1930 of the extreme susceptibility to inbreeding which I have described above as the Achilles heel of the honeybee. Our very reliable isolation apiary, which we have used continuously since 1925, long ago gave us knowledge of problems the existence of which is only now being generally recognised.

In this second part of this book I have to fall back on our own findings. The reason for this is clear: I am unaware of anyone possessing a similar range of experience in the sphere of cross breeding and the formation of new genetic combinations nor anybody with a similar comprehensive knowledge of the various geographical races of the honeybee and their respective characteristics, gained at first-hand. This fund of information has moreover been acquired over a period of close on seventy years by means of an extensive and intensive system of beekeeping, founded on the broadest possible basis. Obviously the key to any successful project in breeding the honeybee lies in a series of accurate findings, confirmed by long experience, and clearly defined objectives. Without these bee breeding is as a ship without a rudder on high seas.

As the subtitle of this book indicates I am dealing here only with the problems which we meet within the breeding of queens. But it must be remembered that the ways and means employed in the rearing of queens can affect essentially certain characteristics of a queen, especially her fecundity, her stamina and in turn the performance of her colony. This does not mean that bad or defective methods of queen rearing have a direct influence on the hereditary factors of a queen but only on their development. I have referred to some of the methods of queen rearing in my book 'Beekeeping at Buckfast Abbey', and I have described there the methods which, in my experience, have proved completely reliable. I make no mention there of the criterion of external markings or of the 'Koersystem' which is the basis of the pure breeding of the Carnica, because these standards are not used anywhere outside German-speaking countries — in fact they are not even known.

Before I proceed further I must say something on a point which I regard as most important, that is the correct use of different technical expressions. The precise use of such expressions is the means of avoiding misunderstandings. Descriptions which are in common use in general breeding of domestic animals do not apply without qualifications in our field. For in the breeding of the honeybee we have to deal with problems which have no counterpart in the breeding of animals. As examples I may mention, parthenogenesis, multiple mating with an unknown number of drones, and also the fact that the millions of sperm in a drone are genetically identical. Such phenomena as these make it impossible for us to compare without qualification the breeding of the honeybee and that of domestic animals. One other factor has to be taken into consideration: the individuals, the queen and the drones, to which reproduction is restricted, give no indication of the characteristics of economic value, (apart from fecundity), which we have to look for in breeding.

In my opinion the following points need to be clearly and precisely defined:

1. Bee breeding is concerned exclusively with the breeding of the honeybee and in no way with the care and maintenance of bees.
2. Pure breeding demands matings within one and the same geographical race and its intensity depends on the degree of inbreeding employed.
3. The term 'Inbreeding' refers to matings of close relatives within the limits of a breeding line, or in a wider sense within a strain or ecotype.
4. The terms 'Crosses' and 'Crossbreeding' refer exclusively to matings between individuals of two or more different geographical races.
5. The terms 'Combinations' and 'Combination breeding' refer to new syntheses of hereditary factors which have been developed from crosses and line breeding over a series of generations.
6. 'Back-crosses' refer to matings effected either between descendants from the original parents or with those of a previous generation of the same cross.
7. 'Mongrels' are the descendants of matings of unknown origin.

It will be seen that such terms as Hybrids, Linehybrids and Hybrid breeding have been avoided. The reason is quite simple: in bee breeding there are no Hybrids. Only a mating between *Apis mellifera* and one of the Indian types of bee would produce a Hybrid, but such a cross is impossible. True Hybrids in the animal and plant world are mostly barren and there can be no reproduction. The classic example of a Hybrid is the mule. Here too effects of heterosis are brought out very clearly, as the mule possesses a stamina and toughness which is not found in its parents, the horse and donkey. In the case of other Hybrids, such as for example a mating of domestic fowl and turkey, fertilisation does take place but there is no normal development of an embryo.

The word Hybrid is of Greek origin and has passed into Latin but always in a derogatory sense. Today in some languages the word is used in a sense of appraisal which contradicts the etymological and natural meaning of the word.

The Aims of Breeding

In his book 'Bienenzuechtungskunde' Armbruster mentions three aims in breeding:

a) a sportive one;

b) a scientific one;

c) an economic one.

Obviously our main concern is with the economic goal of breeding. We are interested in the scientific side only in so far as it can throw light on data which we need for our main purpose. It is, of course, possible to set up a programme of breeding for scientific purposes but even so it should not be pursued without an eye to combining theory and practice. In the first part of this book, the 'Theoretical Part', we drew on science for the basic principles as indicated by the present state of our knowledge. In this second part our emphasis will be on the practical aspects. Although as a practical beekeeper I base my findings on a sound scientific foundation, I am not concerned in these pages with any purely academic goal of breeding.

The Economic Goals in Breeding

From the purely business point of view many beekeepers look for a maximum production of honey per colony as the goal of their endeavours, without taking into account the other factors demanded by our present day circumstances. We do indeed need a high yield per colony but one which must be linked to a minimum expenditure of time and labour. As experience has shown a bee management which goes for high yields of honey without regard to these other needs proves in the end unprofitable. A really profitable beekeeper must combine all the factors I have mentioned in a balanced way, above all with an eye to labour saving. The time factor plays a more important role today than a few more pounds in the crop of honey; a point which must be taken into account in every comparative test.

Whenever a colony records a very high performance we must take it for granted that it was achieved in conjunction with all the other requirements for assessing performance. A colony which is in any way defective be it through disease or a powerful swarming tendency can never show its true performance ability. In all our breeding efforts we must be able to determine all the factors, good and bad, which in any way influence the performance capacity of a colony. Some of these have a bearing on the labour entailed, for example, the swarming tendency; other hereditary characteristics, although not having any direct effect on either the labour or performance aspects, for example, the way the honey cappings are made, do have an economic value and must be given due consideration.

In all our evaluations we have to pay attention to the side effects of certain factors because what we require nearly always depends on a series of genes. It is often assumed that a characteristic depends on one gene alone, but this is not borne out by experience. In almost every case the genes are linked. A comprehensive knowledge and assessment of the characteristics, which we can influence by breeding, forms the necessary foundation for a successful breeding programme. In order to understand the significance of individual characteristics and their mutual influence on each other, for profitable beekeeping, it seems to me appropriate to divide them into three main groups. The first group comprises those primary qualities which are essential for any maximum honey production; the second, qualities which are subsidiary but nevertheless have a direct influence on the performance of a colony; the third group includes those qualities which have a bearing on management especially on reducing the labour involved, and also some others which have but an aesthetic value.

The Primary Qualities for Performance

1. Fecundity
The basic requirement for a colony to be able to produce the very best results is that it should be always at optimum strength at all times. This is effected by a combination of the fecundity of the queen and the industry of the bees in brood rearing, both of which traits are hereditary. Without this optimum strength of a colony the best honey crops are as good as unobtainable.

Great fecundity alone is not the decisive factor but it is the essential prerequisite for any exceptional performance. Some bees, such as the American-Italian strains show an abnormal fecundity and brood rearing tendency. Yet it is always accompanied by a loss of vitality and longevity in the worker bees. This kind of fecundity is obviously no advantage except where bees are reared for sale.

There are essential differences in the fecundity and brood rearing tendency of the different races and strains of bees. The Old-English bee, which belonged to the dark west European race group, even in the most favourable circumstances, hardly ever produced more than eight combs of brood (comb measurement of 19 x 34 cm). This very meagre fecundity was partly compensated by a more than usual vitality and longevity. Yet in spite of this the Old-English bee even in the most propitious conditions never attained anything corresponding to the high averages of crops which today we take for granted. She gathered only a third of the crop made by a prolific Italian colony of that time.

There are without doubt conflicting views about this quality in bee breeding. In England there was at one time the slogan 'We do not want bees, we want honey'. Obviously the commercial beekeeper does not want colonies which turn every pound of honey into brood. But it is the bees who make the honey and the stronger the colony the more likelihood there is of highest production of honey. As I have already mentioned however, an adequate fecundity must be bound up with a series of other essential characteristics of economic value. One good trait demands a whole chain of other good traits. Indeed without the interaction of this chain, one single characteristic cannot develop to the full.

A queen which in the period from the end of May until the end of July cannot maintain with brood nine or ten Dadant combs (46 x 27 cm) does not come up to our standards. This amount of brood must at the same time be spontaneous, by which I mean it must spring naturally from the colony and be sustained without any form of stimulant feeding.

2. Industry or Foraging Zeal

A high degree of fecundity must be accompanied by a boundless capacity for work in foraging. Industry is the lever which raises all the qualities of economic value to our advantage. Although the bee is proverbially busy, there are races and strains which are literally good-for-nothings, a feature which comes out only when comparisons of performance are made in identical conditions. In our isolation apiary where the preliminary selections from the different strains and lines are made, these differences often appear very quickly and always prove well-founded. Industry to a large extent depends on the condition and general state of a colony. A swarm with a newly mated queen, for example, always works for the first few weeks with unparalleled industry.

Although industry is a hereditary characteristic its full development depends on the collaboration of a series of qualities as well as on the actual state of a colony, for example, whether it is queenright or not. From the genetic point of view close inbreeding to intensify this quality of industry can be counter-productive and even bring about a serious deterioration in performance.

3. Resistance to Disease

There can obviously be no hope of maximum performance where a colony is liable to any kind of disease. As with animal and plant breeding so with breeding strains of bees, which in general are very sturdy, this point of resistance to disease is one of our most important considerations. A highly developed resistance to disease is an indispensable factor for successful beekeeping. Ideally, of course, immunity would be the answer, but that is not really possible with bees because we have to deal with a colony which is a community, and not with isolated individuals as is normally the case.

Since we have to deal with a number of diseases both of mature bees and of the brood, and a detailed description of all the factors involved is not possible at this stage, I will return to this subject later in a special section.

4. Disinclination to Swarm

A highly developed disinclination to swarm is an indispensable prerequisite in modern beekeeping. Swarming not only causes untold extra work and loss of time but prevents maximum production per colony, which is precisely the aim of commercial beekeeping in all countries of the world today. A strain of bees which is endowed with every desirable trait but given to abnormal swarming has no place in modern beekeeping. All the good qualities of the strain are lost by swarming. Our experience with the Nigra, to which I shall return later, is a classic instance.

By means of the different swarm-control methods swarming can be made impossible, but such methods as Demaree, artificial swarming or the removal of the queen and making up nuclei are operations which take up so much time that they have no place in commercial beekeeping. Moreover colonies which have been subjected to such treatment are never in a condition to produce maximum crops of honey. Disinclination to swarm, just as the opposite, swarming fever, is hereditary but likewise is dependent on circumstances. Of course both characteristics are subject to a wide range of secondary influences. Inbreeding on a narrow basis reduces the tendency to swarm; whilst crossing, due to heterosis, will intensify the swarming tendency.

The only way to intensify a disinclination to swarm is by breeding. The real problem is to produce a balance between this characteristic and a maximum vitality. Looked at realistically a high degree of disinclination to swarm is preferable to a slight increase in honey production.

Secondary Qualities

Fertility, industry, resistance to disease and disinclination to swarm are the essential qualities of economic importance and form the basis for all our breeding efforts. The other characteristics are not essential in the same way, but they are of very great importance as each one contributes to an intensification of the honey-gathering ability of a colony.

1. Longevity

The quality which must head the list of those which have a very noticeable influence on the performance of a colony is longevity. In fact there is hardly another factor which promises such major possibilities in breeding. There are considerable differences in longevity among the different races and strains of bees. A prolongation of a life-span of a bee, even if only by a few days will bring about a corresponding increase in the number of foraging bees and therefore a higher performance per colony without a corresponding increase in brood rearing. Longevity seems to be due to two factors:
a) an hereditary trait;
b) the quality of nutrition of the larvae during their period of development.
The quality of nursing each individual, be it queen, worker or drone, during their larval development is the crucial point, when it comes to longevity. As far as the queen is concerned it affects her laying capacity as well and in the case of the drone his virility, his ability to reproduce. Inbreeding on the other hand can have a disastrous effect on longevity.

Certain races, notably the Anatolian, Carniolan and West European races are long-lived, a fact which is manifested in the longevity of the queens. Ultra-prolific strains are nearly always short-lived. Exceptional longevity is found more in those races which are moderately prolific, such as the above named. In our experience we have found that bees of Anatolian descent are the most long-lived.

We speak about longevity but it would be more accurate to call what we are concerned with, vitality. After all the length of a bee's life depends on her energy output. The greater the amount of energy expended, the shorter her life-span. Hence longevity in the case of the honeybee is not a matter of predetermined life-span but depends on the store of energy and actual rate of expenditure by each individual. However, the degree of longevity is undoubtedly a genetically determined trait.

2. Wing-power

A highly-developed wing power can materially influence the flight range of a bee. In fact it decides whether or not she can reach a particular source of nectar. I may quote an example from my own experience. Until 1916 when we had the Old-English bee, which shared with the other West European races an extraordinary wing-power, we regularly obtained a crop of heather honey from our home apiary. The nearest heather was some 3.6 km distant at a height of nearly 400 metres above sea-level. In spite of this distance and a rise of close to 400 metres the native bee and her crosses in 1915 made an average of 50 kg of heather honey per colony. Since then only very seldom and then only when the weather was exceptionally good have we had heather honey crops at this home apiary.

On the other hand, the marked wing-power of the western European race group is responsible for the dominant counter-productive influence of the undesirable characteristics, where random matings have to be relied on. Experience has shown this to be so in all those countries where this group is found. Thus a characteristic which is good in itself can be a disadvantage from another point of view.

3. Keen Sense of Smell

Corresponding to an above-average wing-power there has to be a keen sense of smell. Presumably without such a keen sense of smell a bee would not venture on the search for nectar beyond a certain distance. This trait, however, has its drawbacks, for it tends to lead to robbing. The two traits seem to be mutually dependent. A bee which has a highly developed sense of smell seems unable to resist the temptation to rob. In my experience the best honey-gatherers are the first to go robbing. My observations lead me to believe that the two traits are complementary.

4. Instinct for Defence

The most effective remedy against robbing is an acute instinct for self-defence. A highly developed instinct of defence is an essential requirement, and it shows itself most developed in the oriental races. The merciless struggle against the many enemies of bees, about which we in temperate climes have no idea, has doubtless been responsible for this highly developed sense in these races.

Inbreeding can reduce this defence instinct to non-existence, which means that a colony thus handicapped will placidly submit to the robbers without a struggle.

5. Hardiness and Ability to Winter

Hardiness and the ability to winter well are linked to a number of other characteristics. Clearly any bee that quickly gets chilled when collecting water or pollen on sunny but cool spring days cannot be described as hardy. On the other hand, resistance to extremely low temperatures is less important. Good wintering is largely determined by an ability to survive over long periods on inferior stores without a cleansing flight and the reaction of a colony to sudden changes in temperature or to disturbances in general. The Carnica, for example, is inclined to fly on a bright day and a rise of temperature, when our own strain, in identical conditions, will remain completely inactive. Indeed our colonies appear as if dead from the beginning of November to the end of February, or until conditions for a satisfactory cleansing flight have arrived in spring. Any activity in inclement weather brings about a loss of energy in bees to no good purpose whatever, as confirmed in all such instances by practical experience.

6. Spring Development

The next important matter is spring development. I hardly need to emphasise that the way in which bees develop in the spring, whether early or late, depends on a hereditary factor. In my experience which, of course, relates to the conditions prevailing in the south-west of the British Isles, the build-up in spring must occur without any stimulative feeding; it must not start before the weather conditions are favourable; once it has begun it must proceed uninterruptedly despite changes in the weather. The Anatolian bee, even when crossed with other strains, is ideal in this matter. Early breeders expend their stamina by flights in unfavourable weather and waste their energy in endeavours which bring no advantage and often, in fact, are positively harmful. It is well known that early breeders are more susceptible to Nosema than those which start breeding later. The latter nearly always overhaul the early breeders, and moreover at the right point of time, as they have not used up their vitality uselessly and inopportunely. The dwindling of colonies, which is constantly being reported on all sides, is very often the result of a premature spring build-up.

What the modern beekeeper requires is a bee which needs no stimulative feeding but one which builds up in the spring spontaneously, on its own initiative. In this way he is spared all the dangers, the expense and the labour involved in an artificial development. Equally spontaneous must be the ability to maintain a correspondingly high degree of breeding until the end of the summer which guarantees a maximum colony strength of young bees for wintering and spring build-up.

7. Thrift

Frugality or thrift is an important quality closely connected with the seasonal development. Here again we have wide differences between one race and another, and between the different strains. The classic examples of extremes are the Carnica on one side and the Italian on the other. The American-bred Italians are generally unthrifty, while the Anatolian bee is more frugal than the Carnica. There is no doubt in my mind that thrift is a quality very much lacking in modern strains, a fact which leads to great drawbacks in beekeeping on its commercial and practical sides.

8. Instinct for Self Provisioning

This characteristic, closely connected with the previous one, normally manifests itself towards the end of the main flow. Storage of honey in the brood chamber at other times, as often happens with self provisioning types, is a great drawback. The Carniolan is most given to premature brood chamber storage; the Italian the least. A middle between these two extremes answers best to our needs. The assessment of the amount of stores each colony possesses on returning from the heather plays an important part in our breeding endeavours.

9. Arrangement of the Honey Stores

The way the bees arrange their stores of honey is another link in the chain of factors which have a connection with self provisioning. We find that there are two opposite tendencies, storing near the brood nest and storing away from the brood. In our breeding we strive for a tendency to store away from the brood nest at the beginning of the season together with the tendency to store close to the brood at the end of the main flow. Storing away from the brood encourages comb building and also foraging while at the same time acting as a brake on the urge to swarm. An unrestricted brood nest from the end of May till the end of July is an absolute essential in regions where there is a late honey flow. During the heather honey flow the beekeeper has no need for concern, the natural urge for self-preservation in the bees brings about the massing of winter stores in the brood chamber.

The Carniolan tends to store near the brood nest, a trait which approaches extremes towards the end of the summer. The Italian, on the contrary, like most of the other races stands midway between the two extremes. The Caucasian is another example of storing honey close to the brood nest. Moreover this bee stores the newly gathered honey on the least number of combs as possible. The advantage of this method of storing honey is that at the end of the flow or when the flow is suddenly halted you do not find an unwelcome number of partially filled or uncapped cells. This is desirable for preserving the quality of the honey especially in damp climates. The Italian tends to the other extreme. From the breeding point of view what we need is a middle course between the two extremes.

10. Wax Production and Comb Building

An innate desire to build comb is a most important characteristic. Any lack of zeal for comb building nearly always leads to a series of unprofitable activities. There are quite noticeable differences of comb building zeal among the different races. In my experience the Old-English bee was the most zealous comb builder. Not only did she build with an extraordinary speed but also to a perfection which has hardly ever been equalled by any other race. We are fortunate in having been able to preserve that characteristic in our own strain. This is a great advantage for us because almost every year we have to renew all the combs in the supers.

A highly developed zest for building exercises an indirect influence on the honey yield, since a bee which lacks this zest for comb building is more inclined to swarm. In my experience the Carnica lacks this comb building urge which, in all likelihood, is one of the causes of her excessive inclination to swarm. Zeal for building comb undoubtedly encourages industry in general.

Closely connected with the zest for comb building is the inclination to construct drone comb. If expansion falters and there is consequently an interruption in the development of a colony, the result is an increase of drone brood and the rearing of an unwanted number of drones. I am convinced that careful breeding can bring about major improvements in the zest for comb building allied with a curtailing of the inclination to rear drones.

11. Gathering of Pollen

The urge to gather pollen is not the same as the urge to gather nectar. The Carniolans and the Italians do not gather pollen to excess. But in general the West European races are inveterate pollen hoarders. Whereas an Italian strain will rarely gather an excess of pollen even in districts rich in pollen, a dark French bee will even carry pollen through the queen excluder and store it in the supers. This quite phenomenal urge for pollen, hardly ever seen in any other race, is hereditary. In districts which are poor in pollen, or where the pollination of crops is a major consideration, this trait would repay cultivation, more especially where there is a scarcity of pollen in the autumn months, for an insufficiency at this period of the season is widely regarded as a primary cause of Nosema.

12. Tongue-reach

At one time tongue-reach was held to be a factor of great importance especially where red clover was grown. Because of their short tongue-reach the West European races are not able to gather nectar from red clover. The Italian and Carniolan did at one time gather substantial crops from the red clover in this country. However, the cultivation of red clover in all lands is decreasing to such an extent that this source of nectar has no role to play today.

As far as I know the question of tongue-reach plays no part in any other source of nectar. In breeding, attention must be paid to a peculiarity of certain races. At least in those districts where the colour of the honey is highly esteemed. There are races and crosses which are inclined to gather nectar of inferior quality in the same place and under the same honeyflow conditions where, for example, the Italian and Carniolan will make honey of the finest quality — clearly another hereditary factor.

Qualities which Influence Management

We now come to those characteristics which do not have an influence on performance or on production but are nevertheless essential for the realisation of our secondary objectives, namely, those connected with the actual management of the bees. We are here mainly concerned with the traits which lighten the beekeeper's burden while at the same time having an aesthetic value.

1. Good Temper

Although opinions vary widely among beekeepers as to the value of the different characteristics, on one point they seem to be unanimous — their approval of good temper in bees. Bad temper makes the work more arduous, causes an unwarranted loss of time, quite apart from the constant trouble with neighbours. Fortunately good temper is an hereditary disposition which can easily be bred into a strain. Our experience has shown that there are no difficulties in breeding sweet tempered bees in a few generations from crosses with even the most wicked of 'stingers'. Sometimes it happens that in the first cross we find a bad temper trait appears, something which was not apparent in either of the parents. On the other hand the opposite does occur. I know of examples of both: a first cross between Syrian and Buckfast and the Adami bee and Buckfast.

It is not only bad temper that we want to weed out, but also the tendency to attack people and pursue them. The classic example of the propensity to attack people is provided by the West European group of races. It is well known that anyone who comes near to such colonies is in danger of being stung without any provocation on his part. This is a characteristic of the dark West European strains wherever they are found. On the other hand the Eastern races are usually quiet as long as the colony is not disturbed. Should this happen the bees show a propensity to attack and to pursue people to a degree that seems hardly possible. Mention is made of this fact in the Bible.

All strains of the honeybee become irritable in low temperatures. Some of them, such as the Anatolian and Saharan, become very aggressive in the cooler early mornings and late evenings, though they are normally as quiet as any other race. In fact, the Saharan bee in temperatures around 20 degrees Centigrade is as quiet as the Carniolan and a colony can be examined without the use of smoke or any other form of subjugation. The opposite is the case when the temperature is below 15 degrees Centigrade.

2. Calm Behaviour

Two further characteristics which lighten the burden of work are a calm behaviour and the ability to stay fast on the combs. Some races and strains are jittery and quick to fly off the combs, with the result that time is lost when handling the colonies and finding a queen is made more difficult. In the Carniolan we have the perfect example of maximum good temper, calmness and ability to stay on the combs.

3. Disinclination to Propolise

The habit of many races to coat all the inner surfaces of a hive with propolis is one of the most unpleasant traits of the honeybee. This completely unnecessary activity in a modern hive considerably increases the work of the beekeeper.

There are some races, such as the Caucasians in which the urge to propolise is at its maximum, whereas others, such as the Egyptian and true Carniolan, very rarely use it. The Carniolan bees which we had more than 50 years ago, instead of propolis used pure wax. This seems to me to be an hereditary characteristic of the genuine Carniolan. It is one which is sadly completely lacking in the modern strains of the Carniolan.

In my experience it is unusually difficult to breed out or even lessen the urge to propolise. It seems to be dependent on a series of dominant factors while the contrary urge is due to recessive factors. Crosses with the Egyptian bee confirmed this assumption.

4. Brace Comb

The presence of brace comb on the division boards, the top bars and crown boards is something which greatly hinders the work of inspecting the colonies. It is a drawback in nearly all the races to one degree or another, less in the orange-yellow Eastern races such as the Fasciata, Syrian, Cyprian, etc., extreme in the Caucasian and some Anatolian strains. It is almost impossible to open a typical Caucasian colony without a crowbar following a honey flow; every comb in the brood chamber has to be forced free. Brace comb not only makes more work but causes the crushing of numbers of bees, sometimes even of the queen, and also provokes stinging. In nature, for example in a hollow tree trunk, brace comb has a purpose. But in the modern type of hive it is a complete disadvantage. Luckily it is a characteristic which can easily be weeded out through cross breeding in a few generations, often as early as the F2 if the pairing is correct.

5. Cleanliness

There are notable differences between the various races as regards their sense of cleanliness and house keeping abilities. Bees which tolerate the presence of mouldy combs have the sense of hygiene at its lowest level of development. There are such races of bees. It is obvious that resistance against wax moths is dependent on a good house keeping sense. Formerly when we had the Old-English bee, we took for granted the presence of wax moths, just as today we do with West European forms of *Apis m. intermissa*. Since the demise of the Old-English bee I have never seen wax moths in our colonies.

Experiments made in America in the fight to prevent brood disease show clearly that resistance to foul brood, is based to a large degree on a very highly developed sense of cleanliness.

6. Honey Cappings

In countries such as England, where there is a demand for section honey, the art of capping the combs is a factor of great importance. The list of patterns and shapes of comb cappings is almost without end. The Old-English bee offered an unparalleled example of the most perfect and artistic cappings. No other race can show the same form of cappings; they were pure white, raised and dome-shaped, and the outline of each cell was clearly delineated.

The form of the cappings, their design and colour are racially determined. This is so with the way the brood is capped. The genuine Carniolan produces snow-white cappings but they are completely flat and lack any shape; the Italian produces mainly white cappings but they are rather crude in their form; the Anatolian's are also white but the lower third is characteristically greyish, and the outline of each cell is only vague; the majority of the Eastern races cap the honey with dark grey cappings. Wherever white cappings appear, there is always a space between the honey and the wax capping. It is possible by a close study of the basic features of honey cappings to determine to which race a bee belongs, in my opinion, one of the most attractive sides of our breeding work.

The breeding for the kind of cappings which come up to our ideal is beset with many difficulties, as the whole process of capping is dependent on very many factors. At times the ideal appears with the Greek bees, *Apis m. cecropia*, but so far it has not been possible to fix this factor. Here and there we have come across very attractive special patterns, but so far all efforts to breed this into a strain have proved fruitless. However progress has been made in this field, and the attainment of attractive honey cappings remains an integral part of our breeding programme.

7. Sense of Orientation

A highly developed sense of orientation or homing instinct is an invaluable advantage, though less where hives are set out individually or in groups of four, each facing in a different direction. On the other hand, when placed in long rows, all facing in one direction, or, as customary on the Continent, where the hives are set out in blocks, two or three on top of each other or put in a bee house, a keen sense of orientation is of paramount importance. In all such instances a lack of orientation will lead to drifting and the spread of infectious diseases.

A poor sense of direction is also manifested in another sphere, namely in the loss of virgin queens returning from a mating flight. The figures of losses show clearly which races have a highly developed sense of direction. Our experience shows that it is most intensely developed in the Egyptian and Cyprian races. This comes as no surprise to anyone aquainted with the native land of these races. The colonies are still today as they have been from earliest times, set out in clay tubes in rows four or five high, joined so as to form a sort of long wall without any noticeable distinction of entrance — that demands a very high sense of direction. In one instance in our mating station we lost only one out of a batch of 110 Cyprian queens and that was late in the season when losses are always high.

These are the essential qualities which we are looking for in our breeding programme. External features serve as a means of ascertaining the purity of a strain, or, as we shall see later, as a starting point for an intensive selection in cross breeding, but they must never be taken as an infallible indication of a high honey gathering ability. As I have pointed out, performance does not depend on one factor but always on the harmonious interplay of a series of factors. The more perfect this harmony, the greater is the potential for performance.

Breeding as a means of Combating Disease

In the section on 'The Primary Qualities for Performance', one of the four characteristics mentioned was resistance to disease. Since this is one of the most important aspects of breeding, and one which demands further discussion, it seems to me to be the place now to go into a detailed consideration of this question of combating disease through selective breeding.

It is now generally recognised that breeding plays a determining role in the fight against disease in both the animal and plant world. True, our first aim in all breeding work is to raise the level of performance and honey production. However, the best results can be obtained only when they are accompanied by no damage through disease. Today we have at our disposal a bewildering number of prophylactics which are on all sides assumed to be absolutely essential. These means however must be regarded as only extra aids, as so often they bring with them very undesirable consequences.

There is moreover another point: when the use of one of these measures has succeeded in cleaning up a disease, there is always the danger of re-infection. As a top priority of all serious practical breeding there must always be the breeding of a strain genetically immune or resistant to disease.

This, of course, raises the question as to whether this struggle against disease by breeding is a possibility. On all sides this possibility has up to now been seriously doubted, admittedly less so recently than a few years ago. All living creatures are provided with the capacity of protecting themselves against disease. It is this capacity that we can influence by breeding. Here the honeybee is no exception. In her case breeding is not concerned with individuals but with colonies of varying size. Moreover, the strength of a colony during the year is subject to considerable fluctuation. Since all living things are provided with the means of protection against disease, in different degrees, why should the honeybee be an exception? Here practical experience must be the deciding factor. But before a definitive answer can be given, the problems involved must first be set out.

We have at our disposal an almost unending mass of material of a scientific and practical nature on this or that disease, its causes, treatment and prevention. There is however a corresponding dearth of information on the fight against disease by means of breeding and also of the possibilities open to us by this means. This lack of information is easy to explain. Only very few scientific institutes or experimental centres have the resources required for a project of this nature. To obtain positive comparisons, not only is there necessary a large number of colonies, but also a number of reference points which only different races and strains can provide. Where we are dealing with only one particular race or a pure-bred line, the only differences which appear are those of that particular race or line. We cannot obtain any enlightenment on the breeding or profitable possibilities of other lines, crosses or races, above all in the realm of fighting disease. In breeding animals and plants the decisive role in this fight is in every case provided by select racial crosses.

The honeybee is no exception to this although we have to deal with problems not met with in the animal and plant world. In the case of the bee we always have to deal with the community, not with isolated individuals. Although this fact has been mentioned it needs stressing here when we are concerned with breeding as a means to combat disease. As a result of multiple mating each colony is made up of groups of indeterminate numbers of half- sisters and super-sisters. Each group has a common mother but fathers of different provenance with different hereditary characteristics. The number of the individuals of each group of sisters and their influence on the unity of a colony can be very varied. Hence in a colony there is a conglomeration of hereditary dispositions which within prescribed limits, make their appearance and it is these which we have to take into consideration especially their reaction to disease. It is clear from this that the different groups of sister queens will not prove resistant or susceptible to a disease in exactly the same way. This becomes even more clear where random matings take place.

Resistance and Immunity

Before we go on to deal with the different diseases and the possibilities of combating them by means of breeding, I think it is essential to give precise definitions of the words 'resistance' and 'immunity'. The two words are often used indiscriminately and this can lead to misunderstanding. I am of the opinion that it is essential to make a clear distinction between the two words — a distinction, incidentally, which corresponds to reality — otherwise there is the danger of endless confusion and quite false assumptions.

There is scarcely any possibility in modern English usage of mistaking one for the other. The word 'resistance' means a hardiness and an ability to overcome weakness, the concept is that of a struggle between two opponents in which the stronger is the winner. 'Immunity', on the other hand, betokens something much more. It means the complete lack of susceptibility to a disease without any exception. Resistance therefore refers to a midway stage between the two extremes of susceptibility to disease and immunity. In other words, resistance shows itself in varying degrees of intensity between these two extremes, indeed, in certain unfavourable conditions it can break down and fail altogether. This can happen to all living organisms. Immunity can be hereditary but an artificial immunity can also be effected through vaccination. Obviously there is no question of vaccination in the case of the honeybee, but there is probably an innate immunity regarding two diseases, paralysis and sac brood.

The practical beekeeper has to be content with a highly developed resistance allowing the bees to be proof against disease, for complete immunity is unobtainable in our case. But this highly developed resistance fully answers all the needs of modern beekeeping as confirmed by wide practical experience.

Contradictory Reports

As I have already mentioned we are dealing here with problems in which even the possibility of success has been called in doubt, and in which there is little of the nature of positive proof at our disposal. It is a specialised area where completely contradictory reports come from the scientist in the laboratory and from the experience of beekeepers in the field. When this happens and hypothesis and theory are in conflict with practical experience, I am of the opinion that the findings culled from dealing with colonies in normal circumstances must be decisive. Experiments which are made in the laboratory do not always correspond to life in nature and the corresponding reactions to it. This is a fact which is not always recognised and in some cases is ignored. One example can be quoted in confirmation of this: in the laboratory a measure against Varroa was 100% successful, but according to the official report was a complete failure in the apiary. Small scale experiments too, made even in normal natural conditions can be misleading. The practical beekeeper, above all the commercial beekeeper, whose livelihood is at stake, must rely wholly and entirely on concrete results. He cannot afford to make mistakes.

In spite of all these doubts and the conflict between theory and practice in this sphere, I have no hesitation in setting out my experiences about the possibility of breeding as a means of combating disease. I feel I can legitimately do this since my experience in endless attempts to deal with this problem covers a period of more than half a century working with almost every race of bee and a great diversity of crosses.

The daily struggle with harsh reality and the constant contact with bees will inevitably give one an insight into and a knowledge of the honeybee which cannot be acquired in the isolation of a study. However in spite of this wealth of experience, I am very conscious of how inadequate this account is. I can perhaps best describe it as a sort of summary of our present understanding of how breeding can be used in the battle against disease, and at the same time of the possibilities open to us in this field.

Diseases of the Adult Bee

Acarine
The greatest losses ever sustained by beekeepers in any land due to a disease must be attributed to Acarine (*Acarapis woodi*). The first incidents of this disease appeared in 1904 on the Isle of Wight. In the course of a very few years the official figures show that 95% of the colonies of bees in Great Britain had been wiped out through this disease. The epidemic reached its height between 1914 and 1916, and it was thus that by chance from my first

days as a beekeeper I had to grapple with this disease. Up to only a few years ago I was in correspondence with the beekeeper on the Isle of Wight in whose apiary fell the first victims to Acarine.

In 'Bee World' for 1968 appeared my comprehensive report entitled 'Isle of Wight or 'Acarine' Disease: Its Historical and Practical Aspects'. In this article I pointed out that in 1919 the efforts of the Ministry of Agriculture in England to make good the losses of bees were based on the fact that the Italian bees were partially resistant to Acarine. This was ascertained at a time when the cause of the disease was still unknown.

Practical experience soon showed conclusively that there were essential differences between the races as regards susceptibility to the disease. We can thank this fact for the unbroken maintenance of our beekeeping during the critical period of the Acarine epidemic. Yet it was by pure chance that some years later we realised that resistance and susceptibility to Acarine were hereditary factors.

The summer of 1921 was one of the best of the century in southwest England. In the course of this season we had two sister queens, belonging to a series of good breeders, which proved outstanding in their performance. As a consequence we used both of them as breeder queens for the following year. Their progeny showed that in one case they were extremely susceptible and in the other highly resistant to Acarine.

Although this was a clear and even convincing case, nevertheless without further evidence it could not be cited as a proof that resistance or susceptibility to Acarine was an hereditary factor. A proof of this nature obviously demanded a sufficient number of incidents, experiments and comparisons to remove all doubts and uncertainty. A further such incident, one of even greater significance soon followed in the wake of the first.

In 1924 we received two breeding queens for experimental purposes from the beekeeping advisor to the then Ministry of Agriculture who had brought them from North America in the previous summer. The queens belonged to a first class Italian strain which had been developed by a well known firm in the United States and was regarded as one of the best honey-producing strains. It soon became clear to us that this was indeed an outstanding strain in our climatic conditions as well; yet after two years of extensive trials we found that in spite of all its excellent qualities this strain of bee would not prove satisfactory in our environment, for the simple reason that it manifested such a very great susceptibility to Acarine.

Now I come to the interesting point of this particular case. Thirty-two years later we decided to import some queens of the same strain to discover if they still showed the same degree of susceptibility to Acarine. The two queens reached us in the middle of July 1958 and we introduced them to colonies in our home apiary so that we could keep a close watch on them. Their development in the following spring left nothing to be desired, and although both colonies had wintered on only four Dadant combs, by the middle of June 1959 they were both on nine combs of that size. We had it in mind to make use of both for breeding purposes. It was not to be.

Towards the end of July there suddenly appeared massive crawling of the bees from one of the two American queens. To exclude any doubt we sent samples at once to Rothamsted for investigation. The official report confirmed our fears; all the bees were infested with Acarine; there was no trace of Nosema, Amoeba, or any other disease. The massive crawling lasted some days and after losing the greater part of the bees the colony improved slightly before the winter. But as could be foreseen it was doomed to extinction and this soon occurred. In the other colony the disease did not appear until the following spring, but then in an even more intensive form.

One further point must be mentioned: at the time when this massive crawling appeared, that is in July 1959, there were another 38 colonies in the home apiary; not one of these showed the slightest sign of Acarine. Moreover that summer was a good season and our colonies made an average of 172 pounds. Yet the favourable conditions of a good honey flow could not act as a brake on the progress of the Acarine. Favourable conditions can normally do this provided that there is a corresponding resistance. It is without doubt that because these decisive factors are either unknown or not given due consideration that so many contradictory views are expressed about this disease. I could quote a number of similar examples from my experience with susceptibility and resistance to Acarine but it is not really necessary. There is no doubt at all that in the cases cited we had an extreme susceptibility to the disease on the one hand and on the other, a highly developed hereditary resistance to it. At least I do not know of any other explanation as regards this disease.

All my experiments show that it is not a question of immunity but only resistance, a power to withstand disease. As I have pointed out this is sufficient for our needs, as is amply confirmed by our experience in the midst of an Acarine infected area. That the most reliable measure against Acarine is breeding is shown by the fact that over the last 38 years we have never had a case of Acarine in any colonies of our strain, in spite of our environment being such that many bees brought into the area very quickly fall victim to the disease. In brief, any bee which is susceptible to Acarine cannot survive in the climatic conditions of south-west Devon.

It is remarkable that all the strains from North America and those from New Zealand, at least those we have tested, have all proved to be extremely susceptible to Acarine. Our findings have been confirmed over the years in all parts of Great Britain as well as in France, where extensive experiments have been made. Why these particular strains should be so susceptible remains a mystery.

From the breeding point of view it would obviously be a great advantage if we could discover the real cause of resistance and susceptibility. But so far all the investigations made in the scientific institutes of France, Italy and Czechoslovakia have determined only that it is not due to any anatomical abnormality.

The cause of resistance to Acarine is therefore unknown to us. It seems likely however that it is connected with the similar phenomenon of a resistance due to age, by which with bees more than nine days old the Acarine mite cannot enter the breathing tubes. Why this is so is not known, but it may be assumed that one and the same cause lies at the root of both forms of resistance. This would mean that in these highly resistant strains under consideration, the resistance due to age would appear earlier and thus prevent or at least hinder the spread of the disease so that a light outbreak would have no marked effect on the health or performance of a colony. This assumption is a striking one and could throw light on many of the aspects of Acarine and the experiences of dealing with it. One other point should be mentioned here: in general Pathology and Biology the cause of a disease is usually known, but this is not the case with immunity, resistance or susceptibility. So ignorance of the cause of resistance to Acarine is no exceptional case.

Hence in the selection of breeders we have no positive guide lines at our disposal except that which can be obtained from examination under a microscope or from the obvious signs of a disease. Likewise we are not able to ascertain which hereditary factors influence resistance or susceptibility. Breeding therefore can be guided only by a total lack of any symptoms of infection. Progress in breeding can be made most rapidly and reliably, in environments in which susceptibility appears very quickly, as for example in South Devon.

I have discussed the different aspects of the fight against Acarine by means of breeding in detail,
1. because we have clear indications of the possibilities which are at our disposal on this matter;
2. because these possibilities have been widely called in question on many sides in spite of the evidence based on concrete findings over a very long period of time.

Nosema

Another disease which the modern beekeeper sooner or later has to deal with is Nosema. I must say at once as far as I can ascertain there is no obvious resistance to this disease among honeybees except possibly among the Eastern races such as the Egyptian, Syrian and above all the Cyprian.

Today there is hardly a colony completely free of Nosema, but in the majority of colonies there is no real sign of illness. There must therefore be susceptibility and resistance here too. In my opinion what causes resistance to or protection from a Nosema outbreak is always the vitality of the colony. Unfavourable circumstances, bad honey flow conditions, inclement weather and senseless interference with the colonies on the part of the beekeeper — all have an influence on the progress of this disease. But in the final analysis it is vitality which controls the intensity of any outbreak of Nosema and hence its development.

The enormous losses caused by this disease are mainly due to modern methods of beekeeping and especially the narrow-minded concern with pure breeding which pays no attention to the dangers of inbreeding. I cannot subscribe to the view that the main cause of Nosema is a lack of pollen. In south-west Devon there is always in August an abundance of pollen, but in spite of this, cases of Nosema turn up, if the decisive factor, the vitality of the colony is lacking. Of course an optimum vitality is not possible without an adequate diet of pollen.

Unlike the case with Acarine we can observe no special resistance to Nosema in any of the races. Hence there is a marked variation in the degree of susceptibility. The Caucasian is a case of high susceptibility as are also any races which are restless during the winter months and start breeding too early in the season.

In the fight against Nosema through breeding we have above all, as far as is practical, to avoid anything which brings about loss of vitality, and here the first point to consider is the question of inbreeding. In my opinion resistance to

Nosema is not due to any one factor but to a series of factors linked together: in addition to vitality a highly developed zest for gathering pollen — in this connection there are many hereditary variations; a general innate quiescence during the winter and spring so that they are proof against any form of disturbance and finally the disinclination to any premature brood rearing.

It is clear that from the breeding point of view we face a complex problem in the case of Nosema. The honeybee seems to tolerate the presence of this parasite to such an extent that the possibility of a symbiosis cannot be excluded — a sort of 'laissez faire' attitude. In brief, resistance to Nosema depends on a whole combination of circumstances which can definitely be influenced by breeding.

Paralysis

Another disease of adult bees is Paralysis or that form of Paralysis which often occurs when bees collect honeydew from pines. I am not competent to compare the two forms of the disease, but it is a fact that the symptoms of the two are very similar if not identical. All the indications are that we are dealing with a group of diseases which greatly resemble one another and occur throughout the world.

According to the research work done by Dr. L. Bailey of Rothamsted, Paralysis is due to a virus infection which can develop only where hereditary susceptibility is present. Seemingly this virus is present in the majority of colonies and as with Nosema does not usually affect the general health of a colony. It seems further likely that it is a question of a series of viruses which would account for the variety of symptoms appearing in Paralysis and its different forms.

We first came into contact with Paralysis after introducing queens from Northern Italy in 1920 and shortly after from Upper Carniola. In the first case the disease was apparent in all the colonies headed by Italian queens but in a mild form and only during the spring. In one of the colonies from Carniola however, the disease was a virulent form and present throughout the summer. In the progeny of the Carniolan colonies Paralysis was in its extreme form and appeared suddenly at the end of August when the bees were on the heather. Dead and dying bees clogged up the bottom boards and between the combs with all the characteristic symptoms of acute Paralysis. There was no question of any poisoning as they were on the heather.

During the last 30 years similar losses were suffered with importations from the Middle East and North Africa. The only difference has been that the high degree of susceptibility was limited to the spring and that when these colonies were helped with bees resistant to the disease they recovered completely and later did extremely well. During the course of the summer those bees susceptible to Paralysis were free from attack, but then later suffered in the same way as twelve months previously. As long as the same queen was in a colony the sequence always repeated itself. These are clearly examples of extreme cases, but over the years we have had to deal with a whole host of cases of resistance and susceptibility ranging from one extreme to the other. The one thing noticeable was that the highly resistant strains recovered without any special treatment and showed no sign of Paralysis from the end of May until the following spring. For a long time I had been sure that in the case of Paralysis we were dealing with an hereditary susceptibility — I suppose one could say a 'negative hereditary resistance'. At the same time I knew that a colony would never be free of the disease as long as the same queen remained in an infected colony, and also that as soon as a colony had been requeened with one from a highly resistant strain every sign of Paralysis disappeared. Even the most disease infected colonies were cured when a queen from a resistant strain was introduced and her progeny had superseded the susceptible part of the colony. In these cases there was never any recurrence. In view of this fact one can speak of immunity in the fullest sense of the word.

In the light of what we have discovered it is clear that Paralysis can be weeded out by means of breeding. However, one must always take into account the consequences of multiple mating especially where mating takes place without any control over the drones. The offspring of a queen gifted with apparent immunity does not always prove to be 100% immune. Side by side with individuals with complete immunity, others show every shade of degree of resistance.

Apart from obvious instances of Paralysis we have nothing to go by in the breeding for resistance to this disease. Nor do we know of any hereditary factors, as in the case of Nosema, which have a bearing on Paralysis. On the other hand, our experiments have shown that it is possible to raise highly resistant queens from stock susceptible to this disease. Experiments such as this are of course possible only when one has especially good lines or crosses at one's disposal and moreover a wide selection.

I must add that I am not completely convinced about the correctness of the claim made by scientists that the real cause of Paralysis is a specific virus. Time and again we found that as a result of a feed with Fumidil there is a complete curing of Paralysis within a short while. According to medical sources antibiotics have no effect on viruses.

Bacterial Septicaemia

Dr. H. Wille of Liebefeld-Bern has established that in Bacterial Septicaemia we have another group of diseases of the adult bee. I will not deal at length with these except to remark that they do explain puzzling phenomena. Today there is scarcely a colony in which there are not agents which cause Nosema or Paralysis. Moreover a mixture of infections is rather the rule than the exception, similar to the case of mammals where we find in the digestive tract colon bacilli which are not in any special way pathogenic.

Dr. Wille maintains that the most effective way of combating Bacterial Septicaemia lies in correct bee management and breeding.

Diseases of the Brood

Up to now I have dealt with diseases which can afflict the adult bees and in the light of experience indicated the possibilities open to us, by means of breeding, of dealing with them. Now I come to the diseases of the brood. I mentioned at the end of the last section the case of Bacterial Septicaemia which touches on brood disease.

My direct experience with brood diseases is limited in the main to my earliest years as a beekeeper. Before 1916 we could find a case of every kind of brood disease in our apiaries, that is, as long as the native English bee existed. After Acarine had wiped out this variety of the West European races and new blood had been introduced through the Italian, all brood diseases disappeared almost over-night. Since that time brood diseases in South Devon have been almost unknown except where colonies on combs have been brought in from other areas. However in such cases, as soon as disease is diagnosed the authorities step in and carry out the authorised treatment.

The fact that with the demise of the native bee, the different brood diseases also disappeared at the same time, is quite remarkable. This was the case not only in our apiaries but also in the whole of South Devon where previously brood diseases had been endemic. There are places today where the state of affairs is similar to that which was once prevalent here, that is, wherever the West European bee is in general use and also in the native habitat of the Intermissa group. During my journeys I was able to corroborate this again and again. However I do not wish to give the impression that the other races are generally resistant to, let alone immune from, brood disease. There is doubtless a great deal of variety. It is, for instance, not merely by chance that in the native land of the Carnica brood disease is almost unknown. All the other bee diseases are present but brood disease is apparently very rare. Complete immunity however seems possible only in exceptional instances and in all other cases we have but a partial resistance.

American Foulbrood

It is clear from the foregoing that with brood diseases there is as much difference in the range of susceptibility and resistance as in the case of diseases of the adult bee. From the breeding point of view the question arises: how can we in the safest and quickest way in this or that case intensify an hereditary factor of resistance, or in the opposite direction weed out factors which cause susceptibility or influence dispositions prone to disease? As we shall see a high degree of resistance to one disease can mean a high degree of susceptibility to another, especially where insufficient care has been taken in the breeding process.

American foulbrood as its name indicates is a disease which is prevalent in North America, and there is no doubt that this disease is the greatest threat to beekeeping as practised there. Thorough inspection of the colonies and regular renewal of combs are unknown in America. To keep the disease in check experiments were made more than 40 years ago to breed a strain of bee with a high degree of resistance to American foulbrood. These experiments were in the main crowned with success. But it transpired that these highly resistant strains were equally very susceptible to European foulbrood. It seems that the efforts made failed because attention had not been paid to the effects of inbreeding.

These experiments were based on the supposition that the resistance factor is due to a very highly developed propensity for cleanliness. This is undoubtedly true in general, but it is unlikely that this is the only factor in this particular case. The research work done by Dr. W. C. Rothenbukler at the University of Columbus, Ohio, gives support to a hypothesis that nurse bees from a resistant colony are able to cure at least part of the already infected larvae. It is not yet clear how they do this. It has been suggested that resistant colonies produce a more bacteriocidal brood food than the non resistant ones. There is however another possibility, a result of multiple mating, that in one and the same colony the resistant factor is found in individual larvae. This idea cannot be dismissed out of hand: it is a probable hypothesis.

The results of the above mentioned experiments seem to indicate that cleanliness is linked with bad temper. Our findings show that such a generalization does not correspond to the facts. The highly resistant quality of the Carnica is an example.

It is obvious that we cannot breed resistant strains in areas where there is no disease nor where there is no possibility of positive comparisons with other colonies. There is one exception here, that is cleanliness which is one of the most important factors in the battle against foulbrood. But this is a characteristic which plays an important role in the general prevention of disease.

European Foulbrood

In the war against European foulbrood the determining factor — though not the only one — as far as we know is vigour. The above mentioned American experiments and their results make this quite clear. It has also been further confirmed by experiments made in Switzerland. There, European foulbrood, also known as Sourbrood, is perhaps more widespread than in any other country.

In Switzerland there is also a susceptibility connected with the strain of bees which a misguided breeding programme has intensified. My experience shows that the West European group of races is apparently most susceptible to all types of brood disease. In fact, this susceptibility is not confined only to both kinds of foulbrood but is open to every kind of brood disease. This became clear to me during my journeys when I was able to see the bees in their native habitat and receive information direct from the beekeepers themselves.

Sac Brood

We have just mentioned the extreme susceptibility of the West European races to all forms of brood disease. Sac brood is indeed a disease which is virtually limited to this group of races and I have never observed a case in a colony of any other variety. Before 1916 this disease could be found in our apiary, but at all times its occurrence was restricted to colonies of the native black bee, or in subsequent years with imported West European queens. After a queen of another race had been introduced to such stocks all signs of an infection cleared up, without the application of any other remedial measures. So we are dealing in this instance with a racially specific susceptibility, restricted to one particular variety of the honeybee, and an equally racially conditioned resistance — if not of immunity — in all the others. This becomes obvious when we consider that in no case was sac brood apparent following the acceptance of a queen from another race.

Chalkbrood

In Chalkbrood we have to all intents and purposes a similar case to the others. At present this disease is prevalent in all the countries where we have the West European bee. It is especially noticeable where inbreeding has still further intensified the susceptibility. Norway is the classic example of this. But it would be incorrect to conclude that Chalkbrood is restricted to only the West European races. Apparently this disease makes its appearance chiefly where the vitality of the bees has been lowered by inbreeding.

Anomalies of the Brood

At this point I must mention a number of anomalies in the brood which cannot be described as diseases since their primary cause is not due to bacteria, viruses or fungi but to hereditary defects. We come into contact with these anomalies in breeding as well as in the fight against disease. Their appearance and their symptoms can now and again lead to mistaking them for specific diseases. We are in fact dealing with defects or 'lethal factors' in the genes which either partially or completely prevent normal development of the individuals and lead to their early death. Defects of this kind are usually referred to as mutation and occur in all forms of living beings not only in bees.

Sterile eggs are a typical example of this. Queens which have this defect lay normally but none of the eggs develop. This form of sterility is an hereditary defect which appears only in the Intermissa group of races or in strains developed from these.

Our experience shows too that there are some anomalies which appear only under certain conditions and only for a time so that it is just by chance that we notice them. We are here on the threshold of an extremely complex state of affairs, a problem which can appear in so many different forms.

It may be taken for granted that hidden defects of this kind affect susceptibility to brood diseases not only directly but also in so far as they influence the general constitution of the adult bee. It is most probable that they cause a weakening of vitality and thereby a shortening of the lifespan. Whatever the cause, any sign of an anomaly must be given the most careful attention where breeding is involved. On the other hand such deviations have now and again to be temporarily tolerated in accordance with the lesson of the parable of the Tares and the Wheat.

Summary

To the question: Is it possible to combat diseases of the honeybee by means of selective breeding? I am able to give an unqualified affirmative. I can do this because it is an answer based on findings which have been confirmed by an experience going back for more than half a century. We can take it for granted that the honeybee, like any other living creature, was provided by Nature with the means of protecting itself against disease. Indeed in the wild the bee was able to survive more easily than in the conditions of modern beekeeping and modern methods. In the wild, Nature saw to it that there was a rigid rejection of any susceptible individuals and an avoidance of the sources of infection. The efforts of the modern beekeeper tend to lead in the opposite direction.

For a right understanding of the possibilities open to us by breeding as a means for fighting disease I must reiterate and emphasise the clear distinction between immunity and resistance. Immunity means the complete lack of susceptibility to a disease; resistance means the power to withstand to some extent the attack of a disease. It would seem that we can speak of immunity only as regards Paralysis and Sac brood. But, in fact, a high degree of resistance meets all the practical needs of a beekeeper. Climate and honeyflow conditions have an influence on the general health as well as on the proneness to disease in bees, but they have no definite effect when the bees enjoy an extreme resistance or suffer from an equally extreme susceptibility.

The development of immune or highly resistant individuals or strains is one of the essential aims in breeding of plants and domestic animals. We must have a similar objective in the case of the honeybee, as without any doubt we have untold possibilities at our disposal to this end. Indeed, taking the long view we can see that a wide range in selection in breeding can be the solution of the problem of disease. We have at our disposal today a great number of medicaments and prophylactic measures against disease, but they demand a great expenditure of time and money which immunity or high resistance can save us.

This question of the fight against disease by means of selective breeding is to my mind such an important section of this book that I felt it right to reiterate and emphasise the essential points at issue.

Evaluation of Performance

Up to now we have been considering the essential characteristics required in breeding the honeybee. In the light of this knowledge we can now consider the problems associated with an evaluation of performance which in fact forms the basis of success in every sphere of breeding. The plant and animal breeders have however a much simpler task, as they have an almost complete control over environment and food supply, at least in the case of domestic animals. In beebreeding however we have to cope with so many uncertain factors that an exact evaluation becomes that more difficult. For an exact evaluation an absolutely impersonal estimation of the data at hand demands an objectivity for which in the bee world we often seek in vain, yet without it progress seems impossible.

An exact evaluation of performance with bees is, as I have said, a very complicated undertaking. It is relative to so many factors, such as the race, the cross, the strain and line as well as environment and honey flow conditions. The honey flow varies from year to year, from district to district, often within a short distance. When a beekeeper talks about performance and honey crop he has to refer to a definite year and honey flow.

The main sources of error in breeding the honeybee lie in another direction. Comparisons within a particular line or strain will merely establish the relative performance of the colonies of a line, but cannot by itself establish the actual value of performance in relation to other lines. Only by comparing a number of lines, strains and races, in identical conditions, are we able to secure the positive criteria, which will allow us to assess performance on an objective basis. Evaluations on such a basis would moreover demonstrate the actual possibilities of performance the honeybee is capable of and lead to real progress in breeding.

Quite obviously any tests of performance, without positive comparisons secured on a wide and comprehensive scale, assessed against a series of definite checkpoints, would clearly tend to prove fallacious. On the other hand, the more numerous the comparative tests, set against one another, and the more frequently such comparisons are repeated, so much more reliable the progress in breeding.

Our comparative tests for performance carried out annually comprise about 700 colonies and nuclei in a series of out-apiaries, subject to different conditions of honey flow. In every apiary queens of both pure-bred stock and the various crosses are represented. The nuclei are tested at the mating station, situated in the heart of Dartmoor, where the newly mated queens remain from the end of June until March in the following year. The comparative tests are made in two stages. The initial one, of the young queens in the nuclei, covers the period until they are transferred into the honey producing colonies in spring. The more distinctive traits as well as differences of performance may show up by the autumn; their respective wintering ability and thriftiness by the following spring, before the final test in the honey producing colonies is carried out. With regard to this decisive final test I have to draw attention anew to the all important factor, namely, the role the actual capacity of the brood chamber plays. A brood chamber which restricts the maximum laying ability of a queen will inevitably bring about an artificial levelling and uniformity in colony strength and a corresponding reduction in the honey producing ability of such colonies. Obvious differences in performance will nevertheless still manifest themselves, due to variations in the swarming tendency, longevity, industry etc., but due to the all important factor — the reduction in colony strength — a genetically determined maximum performance is effectively excluded; likewise too any criteria by which a true evaluation of the performance of a colony could be assessed. We are in fact deprived of the essential basis of selection and thereby the means of making progress in the breeding of the honeybee.

We regard the preliminary tests carried out in the nuclei, in the heart of Dartmoor, an invaluable help in our evaluations. The nuclei are on four half-Dadant combs and there are four nuclei in each hive with the entrances facing four different ways. The mating station is situated at a height of 1,200 ft., where climatic conditions are unusually harsh throughout the year, with temperatures as low as −18 C in winter. From the middle of October until March the temperature virtually never rises above 10 C. These initial tests are therefore of the severest kind.

Apart from the restriction on the laying potential of queens there is another factor which can render production tests very unreliable and misleading and that is drifting. Where colonies are arranged individually and certain safeguards applied, drifting can be largely avoided. On the other hand, where colonies are set out in rows, all the hives facing in one direction, drifting cannot be avoided, and hence reliable evaluation of performance is equally impossible. An accurate assessment of performance must be free as far as possible of all errors, prejudices and wishful thinking. Success in breeding the honeybee can only be achieved by a completely objective approach. One final remark: the results of comparative tests secured between 10 colonies are obviously much less reliable than say between 100 and still less when 500 are involved. The higher the number of colonies, the more reliable the findings and the greater the chance of progress in breeding.

Breeding Procedures

We have at our disposal a series of breeding procedures and I will now deal with their aims and their respective merits and demerits. I must however return to the point I mentioned previously about the necessity for clearly defined ideas expressed in precise terminology. This is essential for avoiding misconceptions and misunderstandings as an arbitrary use of technical terms is bound to cause confusion.

For breeding the honeybee we have the following procedures at our disposal:
1. Pure breeding
2. Line breeding
3. Cross breeding
4. Combination breeding.

As will soon become clear these four ways are mutually dependent. Pure breeding is the basis of all breeding work; line breeding serves to retain the vitality of pure races and strains; cross breeding of the different races leads to a combining of characteristics with the maximum heterosis effect. These in turn form the bridge to combinations of new hereditary factors which are not otherwise found in nature.

Nature's Method of Breeding

The first point we must deal with is Nature's own breeding method. This is essential because this shows us the right lines we must always keep to in all our procedures. To ignore them inevitably leads sooner or later to failure. Right up to the introduction of modern beekeeping, Nature had been the sole arbiter in the breeding of the honeybee.

In no instance does Nature breed for the highest performance per colony, but only for the preservation of the species and its adaptation to existing circumstances. To preserve the maximum vitality of the bee, Nature introduced a large number of measures to this end. To prevent inbreeding — the Achilles heel of the honeybee — she arranged that mating should take place in free flight up to a distance of 4 miles from the hive and in addition that it should be multiple mating with a number of different drones. Bees were limited to a pattern of pure breeding within a geographical race, a restriction they could not overcome. Nevertheless, Nature made every effort to ensure the preservation of many different characteristics. Any individual that did not measure up to standard was brutally eliminated. Uniformity, whether of external features or of physiological traits, played no part in Nature's design at any time.

Admittedly we have now and again to take measures in our breeding which are not completely consistent with those used by Nature. But on the other hand we have possibilities at our disposal which were denied to her. In the long run we cannot however afford to ignore the guidelines she set out.

Pure Breeding

By means of mating close relations within a particular race or strain we enjoy breeding possibilities which allow us to concentrate, intensify and stabilise the worthwhile characteristics we need. At the same time we are able to weed out step by step the unwanted traits. Without these possibilities no progress could be made in breeding. Mere chance would determine the results. Hence pure breeding is the sheet anchor to keep secure what we obtain and affords that permanence and stability we need in our work. At the same time it provides the firm foundation for successful cross and combination breeding. In fact pure breeding is the bridge essential for any progress in breeding. Up to now uniformity in external characteristics has been given too much importance in breeding experiments; in some quarters it has been given top priority. In my work the opposite is the case. While external characteristics are not to be disregarded, they play only a secondary role.

With the honeybee however there are definite limits to pure breeding. To intensify a particular genetic factor postulates inbreeding. Now the honeybee is very susceptible to the results of inbreeding; the causes of this have already been mentioned. The results of it show themselves in many ways, the worst of which is the loss of vitality. This defect shows up in every phase of the life of the bee but most clearly in its effect on performance. As far back as 1928 I was aware of this susceptibility to the effects of inbreeding, but it was only in recent years that the 'writing on the wall' was widely and correctly interpreted. The results secured by the use of instrumental insemination put the matter beyond any doubt.

Lack of vitality appears most clearly in acute infections of Nosema, a tardiness or complete failure to build up in the spring and a high loss of colonies. The last factor is often ascribed to unfavourable weather conditions whereas in reality the losses are due to the lack of vitality. As we now know, loss of vitality makes survival during the winter and the efficiency of a colony impossible. Inbreeding is without doubt the Achilles heel of the honeybee. The loss of vitality, resulting from close inbreeding, determines the bounds to pure breeding and in the long run we cannot ignore Nature's clear guidelines with impunity. At the same time we have to put up with the disadvantages consequent on inbreeding for the sake of success in certain breeding experiments.

Pure breeding has another disadvantage: it is strictly limited to the characteristics there present and can neither create new ones or new combinations. Such possibilities are exclusively reserved to cross breeding and the synthesization of fixed new combinations.

Line Breeding

To hold and permanently maintain pure stock a carefully considered programme of line breeding presents the only alternative. We have thus maintained our own strain over many years without loss of vigour, stamina and productivity. The lines in question arose from a series of select breeder queens, which had answered our demands in the strictly controlled tests for performance. The individual lines were crossed reciprocally, but not to a fixed preconceived scheme. The matings are in every case arranged to complement each other's characters to the best possible advantage. A series of lines permits such a scheme of planned maintenance of pure stock.

Today the term 'line-matings' is widely used in place of 'line-crosses', 'line-hybrids', 'line-combinations'. This inevitably gives rise to misunderstandings and confusions. The sole use of the term 'line-matings' would eliminate any uncertainty.

Cross Breeding

This obviously refers to matings between different races and as I have already said the full practical advantages of pure breeding are only apparent through cross breeding. This happens in every sphere of breeding be it with animals or plants. When we look around we see at once that the individuals which produce the most are almost without exception crosses, the results of cross breeding. The increase of production in every sphere of agriculture would be impossible without cross breeding. It is the key which opens the door for the best performance and practical success. Beekeeping is no exception to this. Indeed it cannot for ever deny itself the economic advantages and potentialities this method of breeding renders possible, it can in fact do so with less impunity than any other forms of production. This fact is well recognised by every commercial beekeeper.

Pure breeding must always play a decisive role in beekeeping. But it fails, can do no other than fail, where maximum performance is the aim. On the other hand a well chosen cross can in a very short time bring about a quite astonishing increase in performance. As far as the honeybee is concerned, because of her extreme susceptibility to inbreeding, she is by nature directed to mixed mating and cross mating so as to preserve her vitality.

In spite of this, cross breeding is a subject which is taboo in German-speaking lands, although it rouses no misgivings in all other parts of the world. Again and again in German periodicals we meet with such statements as 'success in cross breeding is possible provided it is restricted to first crosses. Further crosses must at all costs be avoided'. This warning has some justification when it is a question of uncontrolled random crosses, but it is quite false when applied to a planned programme of cross breeding.

From time immemorial beekeeping has relied on cross matings. There are a few beekeepers today who are able to rely on pure breeding, but by far the greater number have to deal with mixed matings or crosses and to some extent with crosses between races whether they realise it or not. These cross matings between individuals of the same race or of different races are mostly, if not exclusively, random matings. For a positive study of crossings of races however, a precise knowledge of their origin is imperative, except where merely utility crossings are at issue. Breeding from proven breeders and mating in isolation can provide the beekeeper, without any great expenditure, with all the practical and economic advantages of the pure stock developed at great expense and effort by the specialist.

When we are dealing with crosses, especially crosses between races, we do not get far without an elementary knowledge of the exceptions provided by the honeybee, and also of the influence exercised by heterosis. We cannot here rely on comparisons from the animal and plant world, not even when it comes to practical results. It is however a fact that a properly conducted programme of crosses between races can produce far better results for the beekeeper than is possible in other spheres where crosses are concerned.

Obviously success is not obtained by a purely arbitary crossing. Races and strains have to match and complement one another. This is true in all forms of life and the honeybee is no exception. But in other respects the bee is an exception. For example, it is not a matter of indifference which race provides the father or the mother as is the case with other living beings. The results can be vastly different. Reciprocal crosses are seldom identical. In bees the female influence is the dominant factor. A cross between races which are compatible brings two results: first, there is the union of two, complementary concentrations of characteristics and the possibility of developing genetic combinations of exceptional economic value which otherwise would not be obtainable. Secondly, in this way the best results of heterosis are obtained, a consequence which shows itself in a higher vitality and raised level of performance. Heterosis appears too in line breeding but never in the same marked degree as in crosses between races.

Now we must consider the two decisive factors which affect our work. These are the consequences of heterosis and the dominant role of the mother, which have to be taken into consideration at all times in the breeding of the honeybee. I will first deal with heterosis, for a realistic appreciation of its effects will light up the problem and difficulties of cross breeding. The dominant role played by the mother, as we shall see, determines the choice of the races and the respective parent.

Heterosis will emphasise not only the desirable qualities but also the unwanted ones, especially the swarming tendency. This basic natural urge often dominates all the other factors which have to do with performance. The result is that first crosses, because of their increased vitality, often spend themselves in an inordinate swarming fever. This extreme swarming tendency decreases in the second and succeeding generations, thus allowing the factors for performance to develop to the full. It is taken for granted of course that the selection of breeders continues in the succeeding generations, thus preventing a fall in the level of performance, at least in the usually accepted sense of that phrase. Admittedly there are variations in performance, a fact which one has to deal with in pure breeding, but the over-all results in the F2 and F3 show that the performance level is very much higher than that of the pure lines from which the cross was made. If this were not the case, cross breeding in the honeybee would be a waste of time.

It is clear that bee breeders have overlooked or not paid sufficient attention to the decisive factor of the undesirable effect of heterosis on swarming, an effect which does not appear in general animal breeding. If attention had been paid to this phenomenon, many of the false opinions about crossing of races would have been avoided. This doubtless is the explanation for the disappointments so often encountered in the crossing of races, for few first crossings give practical worth-while results.

Before I can give some classic examples of this dominance and influence of heterosis, I must draw attention to another particularity of the honeybee; it is only in exceptional cases that reciprocal crosses produce the same results. I know of only two such exceptions: the Buckfast and the Greek bee. Both the matings, be it with males or females, produce a progeny which is good-tempered, not given to swarming and capable of good performance.

As far as temper is concerned all crosses between races have a bad reputation. In the majority of cases the blame for this lies with the drones of West European origin. Drones of this lineage always produce an aggressive progeny even when they are crossed with the most gentle races and strains. On the other hand gentle strains crossed with gentle strains can produce an aggressive bee. And yet we can get an extremely good-tempered bee from the worst and meanest of 'stinger'. An increase of the propensity to sting is not a necessary concomitant of crosses between races. As experience shows, by using the dominant character of the mother and by back-crossing, the danger of propensity for stinging can be surmounted. An example: an Anatolian-Buckfast cross produces a bee which is gentle, not given to swarming and one endowed with a high production potential; the cross Buckfast-Anatolian gives a bee equally unwilling to swarm, very productive but very bad tempered.

As I have already indicated the majority of first-crosses are of no practical value because of their extreme proclivity to swarming. There are however exceptions and they can be used as ordinary 'run of the mill' queens for beekeepers who want the advantages of heterosis without great expense and in the easiest possible way.

Our experience has shown that the following first-crosses give the best results without an exceptional tendency to swarm:

Anatolia x Buckfast,
Buckfast x Carnica,
Greek x Buckfast,
Carnica and Sahariensis x Buckfast.

Some first crosses which prove to be of no practical value in the F1, when carefully selected in the F2 and back-crossed (in this case with Buckfast drones) produce quite exceptional results. Again in our experience these are:

French queens x Buckfast,
Swedish and Finnish queens x Buckfast.
Queens mated to Italian or Greek drones would probably give the same results.

Since first-crosses in the above mentioned categories do not prove to be practical for our purposes only a small series is raised in cases of this kind, and only the most suitable which have been through a careful testing are selected for use in the F2. By using such precautionary measures we are able to obtain the fullest practical advantages of these special crosses without undue loss and great expense. Losses in testing such crosses can prove financially very considerable, as the following example shows. In the summer of 1949 we had 30 colonies headed by F1 Nigra-Buckfast queens. To form an unprejudiced judgement on them we distributed them among all our apiaries. 1949 was a very good honey year with an average per colony of 145 pounds. The Nigra first-crosses in identical conditions managed only 22 pounds per colony. This almost incredible difference was due mainly to the extreme swarming tendency of this particular cross. A second cross, even with random mating, was in the following year outstandingly productive. This example shows clearly how first-crosses can prove disappointing, but at the same time how the best advantages of such crosses from the practical point of view can be obtained in the F2, and moreover shows how we can circumvent the disadvantages of the F1.

So far we have mentioned the bad effects of heterosis on the swarming tendency and temper of the bees. Both these characteristics are undesirable species linked traits of the honeybee, which are obviously unknown in general animal and plant breeding. The influence of heterosis is however not restricted to swarming and bad temper; it extends to vitality, resistance to disease, indeed to all the traits which have a bearing on performance. This influence of heterosis manifests itself to the full in the end result, the production of honey. The lack of any one characteristic is mirrored in this end result, but in a negative way. It is only where all the factors which have an influence on production reach their full potential and complement themselves reciprocally that the best results can be obtained.

Here I must emphasise again the importance of one characteristic especially, that is fertility. The performance of any colony is closely bound up with this trait. Indeed fertility plays an important role in cross breeding.

It is widely held that first-crosses are always very fertile, more fertile even than the parent stock. This is only partially true. In my experience heterosis has no noteworthy infuence on the laying ability of a Carnica-Buckfast, Carnica-Italian or a Carnica-Greek first-cross. The brood of these crosses is indeed healthier and more compact but the amount of brood is not noticeably higher. The only exception to this among the Carnica strains is the Sklenar. On the other hand the reciprocal crosses Buckfast-Carnica, Italian-Carnica and above all Greeks in the F1 all show the opposite result, a greatly increased fertility. The classic examples of a marked increase of fertility are Cypria x Buckfast, Cypria x Carnica, Sahariensis x Buckfast and Anatolica x Buckfast first-crosses, and of course in all crosses between races in the F2.

There are many different and contradictory views about the practical value of fertility. But the fact of the matter is that fertility, allied to the climate and honey-flow conditions, is the basis for any successful beekeeping enterprise. From the breeding point of view second-crosses provide us with the best possibilities.

To avoid any misunderstanding I must insist on this point that in all the experiments we made with crosses, the results in every case were obtained through a number of trials and comparisons. To arrive at a completely objective judgement selective crossings are obviously essential, and this was always done. If we felt the need for particular facts and findings we resorted likewise to random mating. In other words we used every possible means to obtain accurate results. Side by side with the results obtained there were of course exceptions, but these exceptions served only to establish the basic rules and norms. The exceptions obviously came by accident, from local conditions and a series of other factors. In beekeeping a highly developed ability or a collection of abilities can easily be a disadvantage. For example, a very prolific and productive bee, which has been placed in a brood chamber which does not correspond to her fecundity, is bound to present a completely false picture of her real capacity and can lead to disappointment about the potential within her. And where this potential cannot develop, the bad points come to the fore. However, the amazing ability of the bee to adapt herself often helps the beekeeper to overcome the results of his ignorance.

Every beekeeper who has to rely on random mating is dealing with crosses, nearly always with crosses of different races, but in any case with mixed matings and drones of mongrel descent. However Nature has put in our hands the means by which we can to a great extent keep within bounds the effects of mixed matings, that is, the dominance of the queen. The influence of the drone in the breeding of the honeybee, both from the theoretical and practical point of view, is far less than in general animal breeding where the opposite is the case. By selecting breeder queens of a particular race we can control the end results of our breeding and also performance. This has always to be taken into consideration when we are dealing with crosses between races; nor must we forget that we can only secure the best possible returns from pure stock when suitably crossed.

These reflections have been set down and the examples quoted so as to give the guide lines for the professional and commercial beekeeper. They are meant to provide him with the information necessary to avoid the pitfalls of cross breeding and also to show him how in the simplest way and with ordinary means he can secure maximum returns from his colonies. In my opinion the problems of cross breeding have up to now been seldom approached from a realistic point of view. The idea that the best results, even in breeding, can be obtained only by complicated methods and procedures does not correspond to reality — in fact, the opposite is the case. It is true that no cross can ever come up to the hopes and expectations of all beekeepers, but this is often due to the fact that many beekeepers cannot bring themselves to make the necessary adaptations and changes of their methods. Scientists have often claimed that crosses can raise average production levels by some 30%. Our experience has shown that a properly conducted breeding programme of crosses can raise the level by as much as 300%. This figure refers to averages not to individual exceptional performances.

Doubtless a claim for an increase in production even by an average of 200% will seem to many a purely Utopian claim. But I think it right to cite a concrete example to show that a claim for a 30% increase comes nowhere near the level which is attainable. The following results are very instructive as they come from three quite different honey seasons.

The year 1956 was for us a season very much below our normal average; the actual average per colony was only 11.7 kg. Yet the Anatolian bees averaged 32.6 kg. The following year was moderately good giving an average of 27 kg per colony, about the same as over the past 40 years, but the Anatolians averaged 64.3 kg. The last example comes from 1964 when we tested a Saharan cross in our apiaries. That summer the general average was 36.6 kg per colony, but that of the Saharans was 113 kg. I should point out that an increase in production of 100% — that is double the average for the apiary — related to an average of 50 kg is of far greater significance than an increase of 30% on an average of only 5 kg. Production figures which do not take into account the apiary averages are worthless.

Combination Breeding

All our experiments with crosses have two aims: first, they are an indispensable means for achieving the maximum yields of honey; secondly, as a preparation for and stepping stone to combination breeding. Cross-breeding alone, although of decided economic value, can produce results of only passing worth and transitory advantage. Periodically the crosses have to be renewed. On the other hand the aim of combination breeding is to fix permanently the advantages obtained so that they have a stability. It is not sufficient to be content with mere transitory gains. The essentially practical advantages of any new combination comes to its peak only through further heterosis, for the greater the possibilities in the parent stock on which the cross is based, so much more intensive is the heterosis. Every combination which keeps a definite goal in view must lead as it always does to a step forward. The honeybee is thus progressively improved at every step and at the same time further possibilities are made available.

I need hardly stress that the term 'new combinations' refers exclusively to syntheses of hereditary factors which are transmitted with the same regularity as in pure stock. A 100% regularity in every factor would of course be possible only in Utopia, but that applies too to any pure stock.

To show that all this is not mere speculation but is all derived from the results of practical experience, I must give a brief account of the history of the Buckfast bee. She owes her origins to a cross made before 1920 between a dark, leather-coloured Italian of that time and the one time English variety of the West European race. About 1940 into this strain a further new combination was introduced, one which had been developed for 10 years from a French cross. Again in 1960 a new combination was incorporated from a Greek cross. These new combinations had been carefully bred for some years and only when the union of the hereditary factors had been definitely fixed and answered all our demands, were they introduced into the main strain. In other words: the new combinations had been bred, over a period of time, tested and step by step 'assembled' before they were finally united with the main stock. The whole purpose of this breeding programme is the attainment of a permanent increase and intensification of the performance potential of the bees. The method adopted is to bring together all the required hereditary factors which exist at their best in the different races. My experience has shown that a minimum of seven years is required for the development of any new combination.

As a pre-requisite for an enterprise of this kind a comprehensive knowledge is necessary of the different races and local types as well as of the individual concentration of particular hereditary factors. We must also know in advance what the breeding possibilities are and how best they are suited to our breeding plans. Each race has its advantages and disadvantages, its good and bad characteristics, always thrown together in different ways and degrees according as chance and circumstances have arbitrarily decreed.

Experience shows that the practical difficulties of breeding new combinations are far from few, but it also shows that the theoretical difficulties have been over estimated. The despair of ever obtaining the ideal, which theoretically is a chance in a million should not worry us unduly. It is not likely that we will obtain the 'jackpot', but nevertheless we can obtain some very profitable new combinations by our efforts. As I have said, these efforts can provide us with new breeds within a comparatively short period of time and certainly quicker than in the breeding of farm stock. In the breeding of the honeybee we can make use of short cuts which save us all the time-consuming and costly efforts required in the breeding of animals.

Development of New Combinations

I have already indicated the essential requirements for the consistent transmission of hereditary factors, namely, the selective crossings of two special races. The choice of these two races is determined by the goal we have in view at the particular moment, or more accurately, the specific factors which we wish to introduce into our strain. As examples we could cite the following: we look for maximum thrift in the Anatolian bee; greatest fertility in the Saharan; reluctance to swarm in the Greek. In each of these cases of course we are not dealing with just one highly developed trait but with an unavoidable series of characteristics. In the course of developing a new combination very often new factors appear of which we previously had no inkling. Now and then these new factors can be undesirable and could even be mutations.

As I have already indicated our first stage in the development programme of new combinations is to test the races in our climatic conditions. The second stage is to cross them experimentally with our own Buckfast bee. These crosses are made on both the mother's and father's side as experience dictates. There are here no hard and fast rules. We must always keep in mind the circumstances and the pro's and con's of each move. With the next stage the turning point is reached in the development of a combination breeding.

In all other forms of breeding a selfing or inbreeding of the F1 brings about the Mendelian segregation with the new combination. As a consequence of parthenogenesis in the honeybee there are no F1 drones in a first-cross, which raises problems for us such as are not encountered in other spheres of breeding. Mendel himself through a selfing of the F1 individuals brought about in the F2 progeny the classic segregation, among them the new synthesis of hereditary factors which bred pure. He therefore had already obtained the hereditary new synthesis in the F2.

On account of parthenogenesis in the honeybee we have to deal with a very complicated state of affairs. An F1 queen produces drones of pure descent. It is only in the F2 queens that we obtain the F1 drones which we need for mating with the daughters of an F1 queen. Thus the required segregation takes place only as a result of an aunt-nephew mating. But here a further problem arises. Since the drones come from unfertilized eggs, the millions of spermatozoa which an individual drone produces are, from the genetic point of view, absolutely identical. In other words: the sons of the F2 queen correspond to the inheritance this queen received from her parents, that is, the grandparents of the drone. Thus we obtain the segregation among the drones of a particular F2 queen, but a complete uniformity in the sperm of each individual drone. As a result of this uniformity in the inheritance of the individual drone there is a greater constancy in the inheritance of the honeybee than in other forms of life. As we now know, the balance is redressed by multiple mating. We can and do obtain the segregation but not in the clear form as in animals and plants where parthenogenesis plays no part.

Intensive Selection

In the honeybee we have therefore the decisive segregation, in which the new combinations of hereditary factors appear, in an aunt-nephew relationship. The segregation is even better in the F3 with certain crosses. But in every case an intensive selection has to be made and followed through. As I have already mentioned, according to theory, millions of individuals are required to attain the ideal combination when a number of hereditary factors is concerned. From this it can be readily seen that only employing the greatest possible number of individuals can lead to success. Clearly we can at best only by chance come across a synthesis of factors which answer our requirements, yet these can lead step by step to the goal we have in view. This is confirmed by experience. The plant breeder is at an advantage here. He knows practically no limits to his selection. But in the case of the honeybee we have only a restricted number of individuals of a few thousand at best.

The intensive selection follows immediately on the emergence of the virgin queens in an incubator. Any other way of this selection is impracticable. Of necessity we have to make this selection according to definite external characteristics; we have no other checkpoints. The number of young queens rejected varies considerably but normally is very high. We count on a loss of 80% in the first selection and a further 10% in the second, which follows the emergence of the first young bees. These last 10% are used to head honey production colonies but will never be utilised for breeding. These are the individuals which do indeed answer our needs from the external characteristics, but are still of mixed hereditary composition. In this way all the undesirable individuals are eliminated.

There is no other way in which we could achieve what we want. The raising of a thousand or so first-rate queens is no problem today by the grafting method and the number of colonies at our disposal. It does not in fact markedly harm the production capacity of our colonies.

As I have said, the selection is based on definite hereditary colour markings. At this stage of the development we have no other criterion by which the selection could be made. These colour markings give us no absolute certainty as to the qualitative and quantitative factors which individuals will eventually manifest. However, they do afford an indication as to what can be expected. As our experiments have clearly shown there is a relationship between colour, physiological factors and behaviour. In this sort of selection experience and judgement play an all-important part. Moreover a sound evaluation demands a strong practical sense, which will not be misled by side-issues or pseudo-scientific considerations.

To assume that with an aunt-nephew pairing we have reached our goal would be false. It is only in the succeeding intensive selection we approach step by step the goal we set ourselves, by a further series of generations, each one subjected to objective testing and selection adapted to each individual need, that we can come with good luck to a fully satisfactory new combination of factors. It would also be wrong to assume that any arbitrary cross could lead to a successful synthesis, there is always the chance that a valuable synthesis can be overlooked or recognised too late. On the other hand an inconspicuous cross can, in the course of its development, produce some extraordinarily good results.

Almost every breeder finds that on occasion he has overestimated the value of extreme colour and uniformity. The breeder of new genetic combinations must also be on guard against such a temptation. Great uniformity is easy to obtain but apparently only at the cost of vitality. I am not aware of any geographical race which exhibits in its homeland the sort of uniformity for which modern beekeepers seek so often. From the commercial point of view such strivings are doomed to be wrecked on the reefs of reality. More than 2,000 years ago Aristotle remarked on the fact that the non-uniform bees of Greece proved better than the uniform type which was there in his time. In everyday practice we have to tolerate, in new combinations, a fairly broad spectrum of variation in colour, as of other external characteristics. What we do demand is a close uniformity in the essential factors of economic importance.

Some Results in Combination Breeding

As in all undertakings, so in this instance, the attained achievements constitute the mark of success. Moreover, the results secured clearly demonstrate the possibilities a purposeful scheme of crossbreeding and the synthesization of genetic combinations can attain in the case of the honeybee. Perhaps I may draw attention to a few of the most notable achievements, for positive results always carry more weight than mere words.

The good temper of the present-day bee is doubtless a classic instance of the progress secured over the years. The temper of the bees of seventy years ago can only be described as 'ferocious'. Good temper is admittedly a trait that has nothing to do with honey gathering dispositions, but it is a characteristic which does lighten our tasks immensely. Indeed, an extremely gentle bee is one of the essentials in modern beekeeping.

Another indispensable trait is a bee that is reluctant to swarm. The customary swarm preventive measures have really no place in a modern apiary. The cost in time and labour cannot be justified economically. I can recall a time when swarms were expected before the end of April; today colonies should show no signs of swarming before the end of June, except in very unusual circumstances.

When they do appear, all that is required as a preventative measure is to tear down the queen cells. However reluctance to swarm is in no way something absolute. Bees which are the most reluctant to swarm will build queen cells if there is an acute shortage of room or they are in any other way restricted. Moreover first-crosses, as a result of heterosis, are always inclined to swarm.

A honey harvest is the outcome of the co-operation of many hereditary factors, and in the final analysis is the decisive element which determines the profitability of any beekeeping. Whereas at one time an average of 4 kg per colony was usual, today we look for averages of more than 40 kg and in good years as much as an average of 100 kg has been obtained. In some cases where bees have been transported to different nectar sources the average has been over 150 kg. These figures are not just for record crops, and they do indicate that quite remarkable increases of production are attainable by means of combination breeding. In fact I see the attainment of exceptional crops as a spur to urge us on to greater efforts in breeding.

A purposeful combination breeding programme must of necessity be orientated by practical and economic considerations; there can be no sideline aims. On the other hand it must not have a mere one-track mentality in the pursuit of a single goal. For instance, it cannot go all-out for a high honey crop production and neglect all the other factors required in beekeeping. What we require is an allround combination, one that is many-sided and breeds true. Without this last mentioned requirement a combination comes to grief. But for any combination or synthesis of this kind which meets every requirement and constantly breeds true to be called a hybrid is a complete misuse of language. True results of lasting value are not acquired over night. They come in stages almost unperceived over a period of years, as is the case with all permanent successes.

Multiple Hybrids

Mention must be made here of the American experiments with the well-known maize hybrids which raised hopes that similar high yields would be achieved with bees. First of all there is no question of these crosses whether with maize or bees being true hybrids: it is rather a matter of line pairings. As a result of this type of breeding the maximum heterosis which is attained in maize, in our experience occurs in the honeybee in only a very modest degree.

This method of breeding involves a very large expenditure, since to obtain the level of heterosis necessary, highly inbred lines procured by means of instrumental insemination through a series of generations are required before line-pairings can be established. The great expense involved is perhaps justified in North America because the required breeding material for genuine crossings of races is not available. Suitable crossings of races bring about more substantial increases in production with minimum expense.

Part III
An Evaluation of the Breeding Possibilities of the Different Races of Honeybees

Introduction

In part I of this treatise I dealt with the theoretical aspect of bee breeding; in part II I discussed the different methods of breeding, their advantages, disadvantages and their possibilities. In part III I would like to turn to an evaluation of the breeding potential of the different geographical races of the honeybee. This part then is concerned with the raw material which Nature has put at our disposal in these different races.

An intelligent breeder must be acquainted in advance with the individual characteristics and capabilities of these races, their pros and cons, something of their makeup, before any breeding programme can be determined. It is similar to the approach of an architect who, for his enterprise to succeed, has to take into consideration the individual characteristics of the materials he intends to use. Bee breeding without a definite plan leads us nowhere. Moreover, unlike the architect, the bee breeder is dealing with living material whose good and bad points cannot be determined with mathematical exactitude.

To evaluate the real worth of the races we have to take into account a very great number of factors for our breeding programme. I would like now to try to describe the results of our experiences with the different races of bees. In order to put this into a concise form, a few repetitions of what I have already said are essential so that we can see how it is that the interplay of all the factors leads to an objective assessment.

Nature's Aims in Breeding

As I have more than once pointed out, from the breeding point of view Nature's sole aim is to preserve and propagate a species in given conditions. She attains her goal by means of a weeding out of any individuals which cannot match up to the demands made on them. In spite of this one-sided selection and limited breeding objective Nature has provided us with a wide range of honeybee races and placed at our disposal a treasury of breeding material of incalculable value.

The notion that a bee native to a particular habitat must of necessity be the best for that region is based on fallacy. In practice experience shows that imports from other areas, even worlds apart, can prove far better than an indigenous variety. For Nature is perforce limited to the characteristics which are present in that race. So, for example, the Carniolan proved better in Central Europe than the native west European race, as likewise does the Italian everywhere.

Acclimatisation

The problem of acclimatisation, which so often is a decisive element in breeding plants and animals, plays no such role in breeding bees. The reason is that the queens and drones on which reproduction depends usually come into contact with the outside world only in the mating flight. A queen passes her life within the hive almost completely isolated from external surroundings. The only variable in her life is the supply of food, but this influences only her laying power.

For long periods of time the bee has adapted herself to her surroundings solely by means of the weeding out of all unsuitable individuals. The experiments made by Dr. J. Louveaux, who transferred strains of bees from the Paris area to southern France and vice versa, are examples of adaptation to local conditions. But this means only that these strains or local types in the given conditions will, when left to themselves, pull through best. Variations of this kind can prove to be advantageous or otherwise, when we exchange certain local types with others or transfer them to districts where there are entirely different surroundings.

The leading honey-producing countries of the world — North and South America, Australia and New Zealand — have no native honeybees, and the importations from Europe have not fared at all badly. They have also produced the clearest proof that there is no such thing as acclimatisation with the honeybee. It is often more difficult for certain races to adapt themselves to rigid types of beekeeping than to a new environment. Importations, if rightly handled, often bring in greater crops of honey than do the so-called acclimatised native bees. Moreover they may survive better hard winters and a difficult spring. In very recent times this has been shown to be true by examples from parts of northern Europe. The experiments too made by Dr. Louveaux show clearly that bees can adapt themselves to the prevailing circumstances. But it would be wrong to conclude from this that because she can adapt herself in this way, only a bee thus conditioned is able to produce a maximum yield in a particular environment. If this were the case, then we should have to rely invariably on strains which have adapted themselves to local surroundings over thousands of years. But this is certainly not the case.

Environment

For a proper understanding of the way we have evaluated the different races our climatic conditions and sources of nectar have to be taken into consideration. Although the essential characteristics of any race do not change as regards their genetic makeup by being placed in a different environment, yet a change of environment does transiently affect the development and moulding of the characteristics of a race. This at the same time brings out their good and bad points.

In Southwest England where all our experiments were conducted we normally have neither the hard winters nor the long hot summers of the Continent. As a result of our location on the southern edge of Dartmoor our average rainfall is 165 cm (65 inches) as against the average for the whole of southern England of some 76 cm (30 inches). The over-riding damp which affects both our winter and summer is not conducive to the best results in beekeeping compared to other part of the British Isles. Total failures are frequent and long rainy periods are a feature of our normal summers. Our main honey flow is from the white clover, *Trifolium repens*, which in good weather yields from the middle of June until the end of July. The heather, *Calluna vulgaris*, provides us with another honey flow from the middle of August until the beginning of September. For this however we have to tranport the colonies on to the Moor. There is a limited source of nectar from the willows, hawthorn and sycamore.

In this sort of climate and in these honey flow conditions we need a bee which is proof against the changeable weather and winter conditions; is resistant to dysentery; is able to build up in spring in spite of the prevailing bad weather; has a highly developed sense of thrift; yet, is at the same time able to maintain an optimal colony strength to make the most of every honey flow. In addition she must be disinclined to swarm, resistant to disease especially Acarine. A bee which is at all inclined to suffer from Nosema, Paralysis or Acarine cannot survive in our part of the world. The year-in, year-out extreme damp together with lack of real warmth and sunshine demands a very vigorous and healthy bee.

From the breeding point of view however these unfavourable climatic conditions have one great advantage; any susceptibility to disease or any other deficiency shows up at once. If the climate and the conditions for honey crops were always favourable, a really reliable evaluation would have a question mark over it, for, as always in nature, in such circumstances the undesirable characteristics and weaknesses do not readily come to the fore. Consistently good crops can, and do, show what a race or cross is capable of, but at the same time they can cover up genetic weaknesses and drawbacks. An unfavourable environment together with great variations in the seasonal crops, interspersed with occasional very good years, provides a more reliable basis for evaluation than would otherwise be the case.

Standards of Reference

Averages of production can be measured in figures. But in our evaluations we take into account factors which are not reflected in crops of honey, which really have nothing to do with honey production and whose range of variations cannot be determined mathematically. For instance, good temper as with many other factors cannot be expressed in figures. In fact, as far as measurement of certain characteristics is concerned we are faced with a big problem in estimating the worth of races of bees. This does not mean that we cannot determine the degree of, for instance, good or bad temper, but we cannot state it in figures.

In any case there must be a standard according to which we can make an assessment of any given race. This is an essential requirement for any objective evaluation of a pure line, a race, a cross or new combination. As far as we are concerned that role is filled by the Buckfast bee. From the commercial point of view the average yields she has made over the years, obtained with a minimum expenditure of time and effort is the determining criterion. This is not always a feature of a race or strain of special value from the breeding point of view. There are certain races and strains which produce quite remarkable crops but at the cost of much time and labour. Yet these strains when suitably crossed may be of great value for breeding.

Biometric Findings

This is perhaps the place to say a word about the importance of biometrics in breeding. These findings give us valuable data about the development and origins of the different races of bees and the relationships between them. However these measurements are concerned only with the external features and cannot give us any direct indication of the physiological traits of the bees or of their behaviour. Thus for example the Carniolan race embraces many local types which in their external features show a great uniformity but in their physiological traits and behaviour are essentially different. The Greek bee, the Cecropia, externally looks very much like the Carniolan, but in almost every other way differs from it as chalk from cheese. However these biometric data provide us with points of reference which form an integral part of modern bee breeding.

My Searches

For close on half a century we have been aware that the only way to make progress in breeding the honeybee was by means of combination breeding, that is, bringing together in one the commercially important characteristics of the various races. These races of the *Apis mellifera* which Nature has handed on to us are widely scattered, and to some extent isolated, mainly in the countries bordering on the Mediterranean. Our first concern therefore, before we could venture on combination breeding on the broadest basis, was to investigate what potential for breeding Nature had given us in the individual races. This meant locating them, collecting them, testing them, evaluating them and finally bringing together the individuals which gave best promise for crossing and combining. Nature could not do this. She left this to the bee breeder of today, but she has over millions of years done the needed spade work for such an undertaking.

Until some 30 years ago all our knowledge of the different races of bees was mostly based on hearsay and suppositions. Now our investigations have given us reliable information about the breeding worth of individual races and local strains, the genetic relationship between the different race groups, their distinguishing morphological and physiological features and extent of their variation. About these decisive details we had previously scant or no appreciation. Yet such exact knowledge can be the only foundation for reliable crossings and combination breeding.

In 1880 the Canadian A. D. Jones, followed two years later by the American Frank Benton, did make an excursion to the Middle East, but their aim was quite different from ours. Both men were looking for a race of bees which would prove better than the Italian. This was obviously a fruitless venture.

The Essential Characteristics of the Races of Bees

Ligustica

This race embraces a number of clearly distinguishable varieties. From the commercial and breeding point of view the best is the dark, leather-coloured bee which has its home in the Ligurian Alps. The lighter coloured variety, which was at one time sent all over the world from the region round Bologna, proved satisfactory everywhere, but showed the drawbacks of the race more clearly than the darker bee. The very light coloured strains, common in North and South America, New Zealand and Australia have a great number of advantages as well as disadvantages but as far as we have ascertained they are less well suited for breeding purposes. The golden variety, or the Aurea, which was once so greatly admired, has proved to be a failure from all practical aspects. Any results which do not state the exact variety in question can lead to mistaken conclusions. The four varieties of the Ligustica cannot be lumped together or brought under one common denominator. Unfortunately this often happens.

From the commercial and breeding point of view the value of the Ligustica lies in a happy synthesis of a great number of good characteristics. Among these we must mention industry, gentleness, fertility, reluctance to swarm, zeal for building comb, white honey-cappings, a willingness to enter supers, cleanliness, resistance to disease, and the tendency to collect flower honey rather than honey dew. The last-named trait is of value only in countries where the colour of the honey determines the price. The Ligustica has shown that she is able to produce good crops from the red clover. In one other characteristic has the Ligustica proved exceptional and that is in her resistance to Acarine. This is especially true of the dark, leather-coloured variety, whereas the golden strains are highly susceptible to Acarine.

However the Ligustica has her drawbacks, and these are serious. She lacks vitality and is inclined to excessive brood rearing. These two faults are the root cause of her other disadvantages. She has too a tendency to drift which is caused by a poor sense of orientation and this can prove a drawback where colonies are set out in rows facing in one direction as is the common practice in apiaries almost world-wide.

Curiously enough, all the above mentioned faults of the Ligustica appear in greatly emphasised form in the very light coloured strains, with an additional one, an unusually high consumption of stores. In European countries such strains have proved highly unsatisfactory as they tend to turn every drop of honey into brood. These light coloured varieties are likewise as already stated unusually susceptible to Acarine. The reason for this is not known in spite of all the work spent on trying to find it. It is all the more surprising when we consider that the dark, leather-coloured Ligustica has over a period of more than 60 years proved to be one of the most resistant to Acarine.

The almost exclusive concentration of these light coloured Italian strains in North America seems to be due to the fact that in sub-tropical Southern and Western States the large queen-rearing centres are concerned mainly with the sale of bees, where honey production plays a secondary role. Hence they need a bee which is given to brood rearing to an extreme degree, something which in entirely different climatic conditions constitutes a serious drawback.

In the dark, leather-coloured Ligustica we have a unique combination of factors of economic and breeding value, thanks to which she has found a welcome in every part of the world. When properly handled she is second to none in answering the needs of the commercial and amateur beekeeper, both when pure bred or crossed. For cross breeding she is suitable both on the mother's and father's side and this applies too when crossed with any other races. This universal aptitude for breeding will establish the Ligustica as the foundation for future developments of combination breeding.

Carnica

We have had this race in our apiaries since the turn of the century. During this period we have tried out no less than sixty different varieties from all parts of Austria, the neighbouring districts of Yugoslavia and Greece, in fact from every country where this bee is found. The fact that we have spread our net so widely for our experiments shows that we had very high hopes for the Carniolan. In Central Europe over the past forty years, she has obtained such a privileged position as to be regarded in some places as 'the best bee'. In England where at one time she was widely favoured, she has now virtually disappeared.

The most important characteristics of the Carniolan are: a remarkably good temper, calmness on the combs, industry, sense of orientation, hardiness and resistance to brood diseases, survival in winter and inclement weather conditions, a highly developed thriftiness, minimum use of propolis, and finally her exceptional tongue-reach which is a great asset where red clover is grown. Her faults are: a premature build-up in spring, reluctance to enter supers and an inclination to suspend brood rearing during pauses in a honey flow.

The most serious drawback in the Carniolan is without doubt her extreme, and very difficult to control, tendency to swarm. Another fault, at least in our conditions, is her susceptibility to Nosema, Paralysis and Acarine. Finally, one which we regard as a serious fault, is her very poor comb building disposition.

We have then in the Carniolan a bee which possesses a long list of valuable traits for breeding purposes, combined with only a few undesirable ones. But, as so often happens, these latter exercise an influence on the valuable ones out of all proportion to their small number. This however is less so from the breeding point of view.

On the breeding side the most valuable features of this bee are her exceptional good temper, her quiet behaviour and steadiness on the combs. On the other hand there are local types in regions where the Carniolan is found which are far from possessing these virtues. Moreover, crosses with the most gentle of Carniolan will now and again give rise to bad tempered offspring. As regards fertility there is a marked difference between the various strains, depending on their places of origin. In our experience we have found none which could fill more than seven Dadant combs with brood. This lack of fertility becomes all the more apparent with their ceasing all brood rearing during a pause in the honey flow. Their snow-white cappings and their use of wax rather than propolis — traits which at one time were greatly prized in England — are now no longer present in any strain of this race, or at least only in a very diminished form. Likewise in the recognised strains of the Carniolan of today we look in vain for the one-time vitality she manifested. The efforts to produce uniformity in the external markings together with the accompanying inbreeding has undoubtedly resulted in a weakening of her vitality.

The quite unusual tendency to swarm which the Carniolan possesses is her most damaging fault. She combines this with the trait of suspending all useful activity while the colony is in the grip of swarming fever. Such an interruption of all useful activities is something we regard as an extreme drawback. First crosses between Carniolan queens and other races always accentuate the swarming tendency and hence are uneconomic. Reciprocal crosses, that is matings with Carniolan drones, are always preferable in our experience.

Although in our conditions we cannot profitably use the Carniolan in preference to other races, I regard her as essential for cross-breeding purposes. In fact, the Carniolan is the key for unlocking the hidden potential of other races, especially the Eastern races. Our experiments have revealed that this bee is something of an enigma. She has enormous potential which is released only in crosses where it can develop. This of course holds good in other races: cross-breeding brings to the fore possibilities which pure bred stock cannot do.

I must mention one point here: where there is a question of a general 'utility' cross with this race of bees, Carniolan drones should be used in every case. A reciprocal cross — Carniolan queens mated with drones of other races — produces very often a bad tempered bee and at the same time a first-cross of little or no economic value. Heterosis intensifies the hereditary swarming tendency to an even greater degree than is normally the case. The result is that such a first-cross expends all its strength in its craze for swarming. Yet in the next and subsequent generations there is a marked decline in the swarming tendency, which allows for the full development of those characteristics which have a direct bearing on honey production, while at the same time there is a greater fecundity, often to a far higher degree than is apparent in the original parent stock.

Sub-varieties of the Carnica

The extent of the native habitat of the Carniolan is very extensive. It comprises the whole of the Balkan Peninsula and adjoining countries in the north. It is not surprising then that in this vast area with all its variations of climate and environment there should be a host of sub-varieties of the Carniolan. Special mention must be made of the Banat bee: those found in the Carpathians, the Pester Plateau in Serbia and those of the Montenegran Alps. Externally all these varieties are hard to distinguish from the typical Carniolan whether from Carniola, Carinthia or Steiermark. However some are less good tempered than others, especially those from the Carpathians. They are on the whole not so prone to swarm, but apart from this they possess no characteristic of any economic value which is not present in a more developed form in the classic Carniolan of north-western Slovenia and Carinthia.

Cecropia

The native bee of Greece undoubtedly belongs to the same family circle as the Carniolan. She is however rightly regarded as a sub-variety, since she differs from the Carniolan in a numbr of important characteristics. But even within the frontiers of Greece itself different strains of this variety can be found. Our experiments show that the kinds found east of the Pindus Mountains from Attica to the northern frontier of the country are the most valuable economically. Seemingly the best variety comes from the Chalkidiki Peninsula due east from Salonika.

Aristotle noted in his time, 2,000 years ago, that the less uniform Greek bee had more vitality than the dark uniform variety, and this is valid today. These bees have neither the light colour nor the uniform appearance, characteristics on which so much value is often placed. Yet in spite of her unattractive external appearance the Greek bee has hardly any equal from the commercial and breeding point of view. As regards temper she is the same as the average Carniolan, but in fecundity she is when crossed unsurpassed by few other races. When crossed with Buckfast drones her reluctance to swarm and the strength of her colonies make her superior to most other races. The strength attained by such colonies is quite phenomenal, yet they do not have any tendency to an excessive brood rearing as is often the case. Contrary to other prolific races the Greek bee has a highly developed housekeeping sense.

Great fertility when not allied with reluctance to swarm would of course be no real advantage, at least in our conditions. A proclivity to swarm renders useless any gain from a more than average fecundity. Put the two desirable characteristics together however, and we have the basis for highly productive and profitable beekeeping. In the fact that these two most important qualities are linked in the Greek bee I see the real value of this race from the breeding point of view. The reluctance to swarm always comes out in first crosses and also in crosses with the Carniolan. Our findings seem to show that the development of a Carniolan strain, markedly reluctant to swarm, seems only possible by way of a cross with a Greek variety.

As regards her less agreeable qualities, the Greek closely resembles the Anatolian, especially in her excessive use of propolis, construction of brace comb and in her watery, flat cappings. But these failings are far less prominent in the Greek; indeed there are some strains where they hardly appear at all. As regards honey cappings we come across occasional incidents of the ideal type of cappings, which we had always regarded as the prerogative of the old English bee. The tendency to build brace comb is easily bred out by crossings. I have found that the Greek bee is very sensitive to inbreeding. On the other hand she is much less susceptible to Nosema, probably because of the great colony strength with which she comes through the winter. I have never noticed any signs of Acarine but some strains have shown indications of Paralysis, especially when there has been inbreeding. The Greek bee comes out best in cross-breeding both on the father's and mother's side. In pure breeding her valuable characteristics do not appear to the full. It is true that there are other races and crosses which have a higher rate of performance, but I have found that this Greek variety is extraordinarily well adapted for combination breeding.

It has always been difficult to obtain first-class stock of this bee for breeding purposes. We imported the first queens in 1952, and we were fortunate with further batches. But in the last twenty years the situation has deteriorated. My fears on this score were already raised in 1952 when I was in the nothern part of the country and saw the accumulation of tens of thousands of colonies brought there from all parts of the land. This is leading to a steady decline anyway of the variety as once found in Macedonia.

Caucasica

We have been experimenting with

Caucasian bees for more than fifty years. The first queens came from North America. In the meantime we have had queens from a number of other reliable sources. Indeed we never had much success with this race. It does seem however from a number of reports that there are strains capable of excellent performance. Yet I wonder sometimes if the comparisons underlying these reports have not been made on too narrow a basis. As we have learned from experience, only a whole series of comparative tests made under identical conditions can provide us with a reliable assessment.

In externals, for example the colour of the grey over-hair, as well as good temper and tongue-reach, this race is very similar to the Carniolan. Against this the Caucasian goes to extremes in building brace-comb and in the use of propolis. As regards these two traits the Caucasian surpasses all other races, although in the use of propolis there are some very close behind her. This abnormal use of propolis and brace-comb makes manipulation of bees in modern hives very difficult, so that in spite of her many good qualities the Caucasian has been restricted from a wider distribution. The extreme use of propolis is transmitted in unmitigated intensity in crosses through a number of generations. Whereas the tendency to build brace-comb is fairly easy to breed out, that for propolising can be eradicated only at the cost of endless trouble. In every race except the Egyptian, this factor is dominant and seems to be due to a great number of alleles.

The Caucasian is universally regarded as the most gentle tempered of races, a fact confirmed by our own experience. However there are strains which are described as genuine Caucasian, but on this point of good temper are far from typical Caucasians. Apart from this trait of good temper the Caucasian is noted for its extreme tongue-reach, which in certain strains of this bee attains the highest average of all races. But it must not be assumed that honey crops from red clover are directly proportionate to tongue-reach; that is, that bees with the greatest tongue-reach infallibly produce the most honey from red clover.

This race is also a classic example of a bee which stores its honey close to the brood. Corresponding to this instinct she stores the honey in a characteristic way on a minimum number of combs. The advantage of this is that at the end of a flow or when a flow suddenly ceases, there are not an undesirable number of partly filled and unsealed combs left on hand. This makes for a better quality of honey especially in a very damp climate. However if these two dispositions are combined with a reluctance in comb building the result is an increase of the tendency to swarm.

As regards fecundity, we have so far discovered no essential difference between this race and the pure Carniolan. The Caucasian reacts in the same way as the Carniolan to a pause in the honey flow; there is a sharp fall in brood rearing. In our climate the Caucasian is highly susceptible to Acarine and Nosema, a fact that is substantiated, at least as regards Nosema, by reports from Central Russia. In general, it is not a race as hardy as would be expected from a mountain bee.

Although this race possesses a number of valuable traits, it is genetically not well-suited for cross-breeding purposes. None of the crosses we have tested have been satisfactory. To sum up: the characteristics of the Caucasian, desirable for our purposes, can be obtained in crosses with other races without the disadvantages and glaring undesirable traits of this race.

Anatolica

Asia Minor and Anatolia is the home not of one race of bees, but of a number of races, and, as one would expect, in the areas where they are intermingled there is a host of intermediary forms. And at the same time we find islands of one race in the middle of an area where another race is widespread. It is often difficult to determine where individuals of a pure strain of a given variety can be found.

The dark bee of the north, that is, the district east of Sinop, shut in between the Black Sea and the Pontus Mountains, differs considerably in its behaviour and its economic qualities from the Caucasian, although there are certain similarities, notably good temper. The orange coloured bee of what was formerly Armenia differs from that of Central Anatolia, which can be regarded as an intermediary form between the other two mentioned races. The bee from Cilicia enclosed in the narrow strip of land between the Taurus range and the Mediterranean and adjoining the Arabian Desert, resembles externally the Syrian bee, and like her, is extremely bad tempered and aggressive. In other respects they are somewhat different. In my experience, the dark grey variety which is predominant in the western part of Asia Minor is the least desirable of all the other varieties domiciled in that country. All these races with the exception of the Syrian, of which only intermediary forms are found within the frontiers of southern Turkey, have certain characteristics in common, although these are very differently emphasised because of the different environment in which the races have their habitat. They are all very thrifty, but as would be expected, the Cilician least of all. For good temper, the Pontus bee leads the way, while the Cilician and the variety from eastern Turkey are at the other end of the scale. But there are strains of both races which can be described as bad tempered or as extremely good tempered. They all, with the exception of the bee from Pontus, have one feature in common, they are sensitive to cold, which shows itself in a marked aggressiveness while such conditions last. This tendency is found in all races but never so marked as in the Anatolian varieties.

In comparison with other races, which we have tried out, the Anatolian varieties with the exception of the Cilician are all below average in fecundity. None of them come up to the Carniolan. However, contrary to the Carniolan, the Anatolians in a first cross are prolific to an almost unbelievable extent, although at the same time all, except those from Central Anatolia, show a marked tendency to swarm. Even in the case of pure lines there are clear differences in the varieties. The Armenian type, for example, when gripped by swarming fever or when they have lost their queen, will build an enormous number of queen cells. Two or three hundred is not uncommon, and in spite of this huge number the young queens are as perfectly developed as they could be, without the slightest sign of undernourishment or any defect.

I have already alluded to their sensitivity to cold, but this shows itself exclusively in an increased stinging propensity and has no bearing on their wintering ability. In fact, as far as wintering is concerned, the Anatolians seem superior to all other races known to me. In the exceptionally cold winter of 1962/63 — coldest in the Southwest of England since 1750 — we wintered in the middle of Dartmoor nuclei of pure Central Anatolian bees on four combs (18.3 x 14.5 cm) with complete success, a feat which seemed scarcely possible in the circumstances. This exceptional ability to winter well, is clearly a result of the extraordinary vitality of the Anatolian which is likewise shown in the longevity of queens and workers. Queens which live up to five years in a large colony are no exceptions. Taking into account their relative fecundity, their exceptional colony strength would be inexplicable without an unusual longevity and powers of endurance. Another remarkable characteristic of this group of races is their highly developed sense of orientation. This shows itself most strikingly in the low loss of queens when returning from mating flights. Over the years we have calculated such losses in our own strain as about 22½%; in the Carniolan 10%; but only 5% in the Anatolian and Cyprian. It may be safely assumed that this characteristic is not restricted only to the queens.

As with every colony or every race in performance or honey-gathering ability a whole chain of factors is reflected. It is not alone the zest for gathering nectar which is the decisive element. The Anatolian varieties possess a synthesis of good factors such as is hardly found in any other race. But among the varieties there is a marked difference in their performance, especially in the first crosses, due partly to the difference of swarming tendency. By far the best results have been obtained from the variety which is found in Central Anatolia, north and north east of Ankara. This variety is moreover endowed with another most valuable trait, that is, frugality. In my findings no other race can compare with her on this point.

The Anatolian group of races has of course its drawbacks. Apart from the bad temper and swarming tendency to which I have already referred, they build a great deal of brace comb and make an inordinate use of propolis. But these defects are by no means so marked as in the Caucasian. In fact when they are crossed these defects appear in a very mitigated form and with careful selection disappear in a few generations. This race also tends to be susceptible to Paralysis, as well as a failure to ripen fully the nectar from *Calluna vulgaris*. The result is that in certain years the heather honey starts to ferment a few days after it has been capped. However our experience has shown these defects, apart from the use of propolis, can be easily eliminated by careful selection.

It is clear that we are dealing here with a group of races, and hence one name and description fails to cover all adequately. Granted that there is a close inter-relationship, the differences between the races of the group are very striking. And by differences I do not mean merely external markings but differences which affect behaviour and physiological characteristics, that is differences of fundamental importance.

Our comparative tests have shown that while the other Anatolians in certain features are superior to those of Central Anatolia, generally speaking, these Central Anatolian bees are economically the best and the most valuable for breeding purposes. We have limited our experiments mainly to crosses with the Buckfast bee, that is, Anatolian queens with Buckfast drones. The reciprocal cross is likewise good in performance but very bad tempered. Heterosis in these two crosses does not accentuate the swarming tendency and the Anatolian-Buckfast cross is much more prolific.

Again I must emphasise that with the pure Central Anatolian bee, as indeed with this whole group of races, it would be futile to expect maximum performance. The really economically valuable characteristics manifest themselves to the full only when suitably crossed. In our evaluation we have only considered selected crosses, never random ones. An unsuitable crossing can produce an extremely bad tempered progeny.

From what I have discovered in our experiments, the Anatolian group of races, especially the Central Anatolian variety, is going to play an all important role in the development of new combinations. In this variety we have at our disposal a linkage of factors of the highest breeding value, perhaps more than in any other race. The few defects it suffers from are no real obstacle to the development of highly satisfactory combinations, as confirmed by our own findings.

Fasciata Race Group

As far as I can determine we have in the Egyptian bee one of the primary races from which sprang the orange coloured group of the Middle East, that is, the Syrian, Cyprian, Cilician and probably the one I have called Armenian. At all events the influence of the Syrian is clearly seen in many parts of Armenia.

The name of the Egyptian bee was changed some years ago to *Apis mellifera Lamarckii*, because Linné had given the name *fasciata* to another insect. I am however keeping to that name because the Egyptian bee has long been known by it. There is no likelihood of a misunderstanding.

Fasciata

In the Egyptian bee we have a race of exceptional uniformity and distinctiveness. It was probably restricted from the earliest times to the area of the Nile Valley and Nile Delta and thus almost completely shut off from the outside world. Thus all the conditions were there to produce an exceptional uniformity. The Egyptian bee is a pretty, charming creature. The bright orange colour of the chitin together with the nearly white pubescence, which makes the bee appear to have been dusted with flour, gives it an irresistible charm. The bright orange extends to the fourth dorsal segment. The ventral segments are almost completely yellow, with the exception of the last two which are dark. The thorax is jet black, as are the dark part of the dorsal segments. The scutellum of the workers is bright orange but that of the queen and drones is black. The abdomen of the queen is a bright orange with a narrow sharply-defined crescent-shaped rim to each segment — the characteristic marking of all oriental races.

The pure Egyptian bee has a moderate fertility, is not particularly given to swarming but is inclined to be aggressive. She does not form a winter cluster. When in a swarming mood they build a very large number of queen cells, not just singly but in clusters even on the face of combs with sealed brood, a characteristic I have not observed in any other race. The queen cells are small and almost smooth. Cappings are very much darker than in any other race. As far as breeding is concerned, the Egyptian has one great advantage: she is the only race of the honeybee which does not make use of propolis, a rare quality she shares with the Indian species. Other highly desirable qualities are her highly developed instinct for self-defence and disinclination to drift.

From the commercial point of view the

Egyptian bee has no noteworthy significance but from the breeding point of view she is of immeasurable value as we have discovered in the Egyptian-Buckfast crosses. The F1 is usually bad tempered, which is to be expected. The next generations however are progressively more gentle, very prolific and unusually calm during manipulation. Queens which are descended from such crosses often go on laying without a break even during observation. I have never seen any other cross-bred stock behave so calmly. However Egyptian crosses have one disadvantage, which is most noticeable in the F1, that is they have no resistance to cold or low temperatures. We can however breed out this deficiency in stages. But, as I have said, other than the Indian species, the Egyptian is the one race of honeybee which does not gather propolis. This is something we value very highly. The gathering of propolis is subject to a number of dominant genetic factors, and hence this characteristic of the Egyptian bee not to propolise is difficult to isolate in new combinations.

Syriaca

The Syrian and the Cyprian varieties are often regarded as being of one race. True, they have many characteristics in common, both good and bad, but there are marked differences, and for anyone with experience this is easily seen. It is obvious that they are closely related. It seems more than likely that the varieties in Anatolia, in the districts south and north east of the Taurus Mountains, as well as those on the border with Mesopotamia, are descended from the Syrian. The Syrian in its turn is an intermediary from between these two races and the Egyptian.

The Syrian bee is a beautiful, attractive bee. In size, colour, the whiteness of its hair, its sensitivity to cold and in other respects it is very close to the Egyptian. This shows itself most strikingly, apart from size and colour, in its sensitivity to cold. The Syrian becomes numb at temperatures at which the Cyprian bee is still very active. As is to be expected this sensitivity has an adverse effect on industry and performance.

The Syrian can be extremely bad tempered. This however is not an unprovoked aggressiveness; but if she is disturbed and her temper aroused, then her ferocity and power of pursuit is enormous. It is not just a matter of individual bees but of literally thousands who will follow one to a great distance from their home. This extremely unpleasant characteristic is a feature of the whole race group, but reaches its greatest intensity in the Syrian and Cyprian.

Since the Syrian is of no real economic value in her homeland, still less has she in other countries. From the breeding point of view, I see no advantages in her nor much possibility of developments. The good qualities which she has are found in a much more useful form and in a greater intensity in the Egyptian, Cyprian and Cilician.

Cypria

My experience with the Cyprian bee extends over a period of more than 70 years. It was not until 1920 however that the first queens direct from Cyprus reached us. They came from the district round Nicosia. In the following year we tried out more than 100 colonies headed by queens of this race crossed with Italian drones. 1921 happened to be a very good honey year, and the good and bad traits of this bee soon became apparent. From then on we imported queens from different parts of the island, and in May 1952 I had an opportunity of seeing this bee in her native habitat. In Northern Europe, as elsewhere, the pure Cyprian has no claim to any economic value. She possesses only a limited fertility, which doubtless answers her needs in her sub-tropical homeland. But when she is suitably crossed, this bee manifests an extraordinary fertility. The industry of such crosses is quite unique. The pure Cyprian is not inclined to swarm, but the first cross is very prone to swarm, no matter with which race it is crossed. However, should a honey flow occur, the swarming fever disappears overnight and the Cyprian can produce exceptionally good crops of honey. The honour of the highest individual performance goes to a Cyprian-Carniolan cross.

The Cyprian possesses a series of exceptionally valuable characteristics but at the same time a number of extremely bad ones. From the commercial and breeding points of view the valuable traits appear only in cross-breeding. One of her most striking characteristics is her ability to winter better than any other race. This is true even in our northern climate in spite of the fact that her native home is in a sub-tropical zone. I have never seen a Cyprian colony, whether pure or crossed, which did not come through winter in perfect condition and which did not surpass all other races and crosses in the spring build-up. This is obviously a result of the immense vitality which this bee possesses. However an extreme vitality of this kind has its drawbacks in other directions.

Nothing has brought this race more into disfavour than its excitable temperament. And, rightly so, as the majority of crosses and strains, especially in cold or unfavourable weather, react to any interference with unparalleled fury. Its stinging propensity is not limited to dealing with any disturbance in the neighbourhood of the hive but extends to a merciless pursuit of the intruder for a considerable distance. This characteristic she shares with the Syrian bee. This extremely undesirable trait appears only when there has been interference with or disturbance of the colony. In Cyprus and Syria I often saw large numbers of primitive hives in small gardens and courtyards, surrounded by houses, where people were continually passing without anyone being molested by the bees. This is a clear indication that this bee is not given to unprovoked attack as, for example, is the common black bee of Western Europe. As can be concluded from its ability to winter well and build up quickly in the spring, the Cyprian bee is particularly resistant to disease, at least to those diseases which affect the adult bees. On this point we have here a bee which has no equal nor have I ever seen any defects in their brood. The Cyprian bee has another undesirable trait; within a very short time of the loss of the queen laying workers make their appearance. This is a tendency of the pure Cyprian as well as the first crosses, a defect she shares with her near-relative, the Syrian bee.

There are some other characteristics possessed by the Cyprian in a very noticeable way which I have to mention. When dealing with the Caucasian and Anatolian group of races I deplored their inclination to build brace comb. The majority of races show signs of this, although some often only sparingly. Brace comb and burr comb can make the handling of a colony difficult and unpleasant. Now in the Cyprian we have a bee which shows no sign of this tendency.

This trait has of course no influence on the honey crop, but it is nevertheless very advantageous on the practical side.

The Cyprian possesses an unsurpassed sense of orientation: we have abundant evidence of this from the low loss of queens in their mating flights. A highly-developed sense of smell is doubtless a necessary pre-requisite for an above-average sense of direction.

The two traits are complementary. The traditional arrangement of the primitive tubular hives in Cyprus is to place them in four or five layers, one on top of the other, in stacks of great length with scarcely any kind of distinguishing mark. Such an arrangement necessitates a faultless sense of orientation and recognition. But as in similar cases, a keen sense of smell has its disadvantages. Bees with this sense are generally difficult to unite. Our tests have shown that these traits are not the sole prerogative of the Cyprian bee but are equally a mark of the whole Fasciata group.

The many valuable characteristics of the Cyprian bee appear at their best only in cross-breeding. A hundred thousand years of inbreeding, within the limits of comparatively few colonies, have covered up the full potentialities of this race. The complete isolation of the island, the inbreeding over thousands of years, the hard living conditions, the scanty resources, the merciless natural selection have conspired together to give us a bee of inestimable value for breeding. But the Cyprian bee is not one for the ordinary commercial beekeeper.

Adami

My research journeys led me to the island of Crete. As I was soon to discover, the Cretan bee is noted for its extreme aggressiveness. Because of this I did not think any advantage would be gained by using them in our experiments. However I collected a number of examples for biometric study. Professor Ruttner carried out this study and was able to determine that here, unexpectedly, was a clearly defined, independent race which up to then had been unknown. He named it *Apis mellifera Adami*.

As far as size and tongue-reach is concerned the Cretan bee belongs to the medium sized bee races, and also to the long-legged group. She has an unusually broad tomentum similar to the Carniolan, and a cubital index as small as any of the western European bees. She has small wings and a broad abdomen with three orange coloured segments and, in contrast to the Cyprian, a dark scutellum. With some other differences she is in marked contrast to the honeybees of the Balkan Peninsula and to the common European varieties. But the characteristics in general point to a near relationship to the bees of Asia Minor and Cyprus.

When these findings had been published, I thought it right to bring this race into our experiments. Their unusual ferocity was soon apparent. At the same time a number of characteristics appeared such as I had never seen in other races. If they lose their queen, they build a great profusion of queen cells in compact clusters on the worker brood, with cappings closely resembling those of drone brood. The individual queen cells are likewise produced in great numbers resembling in size and shape those of the Egyptian bee.

To my great surprise the F1 crossed with drones from our own strain turned out to be almost as quiet as the pure Buckfast, extraordinarily prolific, not given to swarming and thrifty. In the building of brace comb it came close to the Caucasian, but not in the use of propolis. As with all crosses, the disadvantageous traits appeared in a diminished form in the successive generations.

Our findings point to a new variety which, when suitably crossed, is capable of very high production and for breeding purposes will be very valuable. Professor Ruttner stated that the biometric results seemed to indicate that this Cretan variety had similarities to Anatolian and Cyprian bees. Our results from experiments in pure breeding and crosses, limited to the physiological traits and the behaviour of the bees, give ground for stating that the Cretan bee is a distant member of the Egyptian race group.

The discovery of this race and the recognition of its commercial and breeding value show once more that only unprejudiced research can determine the real value of any variety of the honeybee. Superficial impressions and conclusions can lead only to completely false evaluations. Testing on the broadest possible basis is in every case an essential prerequisite for a genuine appreciation.

Intermissa Race Group

Intermissa

The native bee of Tunis, Algeria and Morocco is another of the primary races. The economic value of this jet black race is very small because of the unusual number of disadvantageous characteristics it possesses. However our experiments have shown that there are definite possibilities in the Intermissa for breeding purposes, not indeed in line breeding but exclusively in cross-breeding and then only in suitable crossings.

The genetic makeup of this bee offers great possibilities, for better and for worse. The good ones are usually masked by the bad ones and they come out only in the progeny of an F1. Her worst faults are her bad temper and nervousness; her tendency to swarm and her inordinate brood rearing; her proneness to disease, especially of the brood, and her tendency to propolise. These faults are found in all the sub-varieties of this race. The eastern races are only aggressive when they are disturbed, but the Intermissa is a 'stinger' by nature, and will attack any mortal thing which comes near her home. She goes to extremes in her nervousness and tendency to fly off the combs, and this holds too for her swarming and excessive brood rearing. All this is intensified of course in the F1. These unfortunate tendencies are still present as late as the end of September at a time when other races and crosses have hardly any brood at all. I have now and again been forced to take away the queen from colonies of this race while we were feeding them for winter so as to prevent the stores being turned into brood. As far as propolis is concerned, not only every bit of wood on the inside of the hive and the frames but also the combs themselves are coated with it. The honey cappings are always dark grey.

This race and its sub-varieties labour furthermore under a very serious hereditary weakness. In the case of the Carniolan I drew attention to its remarkable resistance to brood diseases. In the Intermissa we have the other extreme: there is a quite extraordinary susceptibility to diseases and defects of the brood. This is clearly noticeable in the native habitat of this race. Brood diseases constitute the chief obstacle to any profitable beekeeping there. We know that these diseases of the honeybee are due in large measure to a lack of vitality caused mainly by a too close inbreeding. But in the Intermissa the susceptibility is caused by a definite genetic defect. Against diseases which affect the adult bees, the Intermissa is very resistant with one exception. This is her extraordinary susceptibility to Acarine. I have never seen a trace of Paralysis.

When dealing with the Anatolian race group, I mentioned the curious inability of a number of races to ripen nectar from *Calluna vulgaris*, at least in certain climatic conditions. I have never observed this unfortunate defect in the Intermissa or any of its sub-varieties. It has, however, been noted in the Norwegian variety.

One other characteristic of the Intermissa which deserves notice is her highly developed urge for gathering pollen. She is really a glutton for pollen. This trait is one which is predominant in all the sub-varieties which owe their origin to the Intermissa. The yellow races gather nothing like the stocks of pollen as do the Intermissa races.

There is no moderation or middle course with the Intermissa, she goes to extremes in everything, a carefree spendthrift, a child of the wild, and yet gifted with the primitive exuberance of vitality and energy. It is the task of modern bee breeding to harness the primitive vitality hidden in this bee and bring its good characteristics to serve our purposes.

Sub-varieties of the Intermissa

As I have already indicated, we have in the North African bee a primary race of which the numerous sub-varieties have spread via the Iberian Peninsula, to Central Europe and Northern Asia as far as the Pacific Ocean. Anyone who is acquainted with the primary race and the Western and Northern European varieties can easily trace these characteristics right down the line. As is to be expected, all these characteristics both economic and uneconomic are found in their highest intensity in the parent race. On the other hand, we find in the Western and Northern European varieties a progressive graduation of some of the Intermissa traits, although they are all present even in the most remote varieties.

The oldest sub-variety is to be found in the Iberian Peninsula where for millions of years it has been domiciled and confined during the Ice Age. Only after the last Ice Age, some 10,000 years ago, was she able to spread northwards around the Pyrenees and gradually find a home in Western Europe and Northern Asia. Obviously when we speak about French, English, Dutch, Swedish, Finnish, Polish and German bees we are talking in very broad terms as bees do not recognise any frontiers. There would be no point moreover in describing these local types individually, as they all without exception possess the essential characteristics of the

Intermissa. There are of course differences between them but these are mainly a matter of the degree in which a particular characteristic of the parent stock is manifested. The jet black colour of the Intermissa comes out especially in the Swiss variety whereas in other countries the colour is brown. Again the extreme tendency to swarm in the Intermissa shows up in the Dutch bee *M. Iehzeni*. One feature all the varieties have in common — unprovoked aggression.

We have made extensive tests of all the above mentioned sub-varieties as well as of the Intermissa itself. In every case what we find are just the essential characteristics of the prototype in varying degrees of intensity. I must however mention one especial feature of the Intermissa race, one which is of very great value both economically and from the breeding side, that is, its ability to build up in the spring from a mere handful of bees to a colony of great strength. I know of no other race which has this capacity to the same degree. We can speak of the 'explosive' development of the Carniolan, but there is really no comparison between it and that of some of the varieties of the Western European races. The development in the latter does not occur so early as with the Carniolan, but when it does come it is much more reliable.

Other worthwhile traits of the varieties of the Intermissa are their wing-power, longevity, industry in comb building and their white cappings of the honey. South of the Pyrenees the cappings are usually dark to very dark. North of the Pyrenees there is a gradual change to lighter cappings to those of the former English bee which produced the perfect cappings.

The Intermissa race group possesses a series of valuable characteristics from the breeding standpoint, but at the same time an equally long series of disadvantages. But in Nature gold is always found among scree and rubble. And we have in this race, as our experience has shown, something of lasting value: it is admirably suited for crosses and combination breeding. The Buckfast bee which we developed from one such cross is a classic example.

Northern European and North Asia Sub-varieties

As I have already remarked, the sub-varieties of the Intermissa race group extend from the far north, south of the Arctic Circle, from the Atlantic to the Pacific. In this immense region there must of necessity be strains of bees with the ability to withstand extremes of cold in winter without any possibility of a flight for periods of months. It is not a matter of just mere hardiness but of a stamina which can survive long periods of confinement without a cleansing flight. This ability brings with it a number of other characteristics. Longevity, a quiet disposition, low consumption of stores and a capacity to subsist on stores of poor quality are among such traits. On the other hand the short summers, limited to but a few months, demand an ability to build up quickly and a swarming tendency to make up for the inevitable winter losses. Any colony which is not in the best condition in autumn is doomed to extinction in Arctic environments. Natural selection here operates in its most savage form.

Following from this we must conclude that the qualities responsible for survival in such extreme circumstances are here found in maximum concentration and firmly anchored genetically. Our findings confirmed this assumption beyond any doubt.

We were well aware of the difficulties involved in isolating these qualities before we made our first experiment in 1968 with a sub-variety from Finland and before that with one from Sweden. As we feared all the undesirable traits of the Intermissa race group appeared in these sub-varieties in a highly intensive form. Although we have worked for some time on isolating the desirable characteristics of these sub-varieties, complete success has so far eluded us.

Apis Mellifera Major Nova

This name is given to another sub-variety of the Intermissa which Professor Ruttner discovered some years ago by chance in the middle of the native habitat of the Intermissa, in the Rif Mountains of Morocco. As a result of his recommendation we visited this area in the spring of 1976, which gave us the opportunity of trying out this special variety.

Our findings show that apart from the size of her body, her tongue-reach and wingspan, her physiological characteristics and her behaviour are identical to those of the Intermissa. She is the largest of any of the honeybees known to us. Her tongue-reach is comparable to that of the Caucasian which means that she has the longest tongue-reach of any known honeybee. The same is true of her wingspan. The Rif bee is however not jet black like the Intermissa but brownish as many of the Northern European varieties. I observed this shade of colour also in other parts of Morocco.

This variety of the Intermissa is indeed a very interesting one. The whole Intermissa race group has a short tongue-reach and a small cubital index. The Rif bee shows the exact opposite. We may presume that in this as yet unexplored region, where over thousands of years different varieties of races have developed, there are more discoveries awaiting us.

According to our findings the Rif bee both pure bred and when crossed manifests a phenomenally high consumption of stores in winter. In identical conditions the Rif colonies averaged a consumption of 14.4 kg, whereas the Anatolian averaged only 6.75 kg and the other strains and crosses 9.45 kg. This high consumption is probably due to its restless behaviour during the winter months.

Sahariensis

Finally we come to a most interesting race, one which has enjoyed an existence for thousands of years in the seclusion of the Moroccan oases of the Sahara. There are oases east of the Moroccan frontier where this bee can be found, but nowhere as I have been able to ascertain, east of Laghouat. It is only a short time ago that even the existence of this bee was doubted. Even today her origins and ancestry are uncertain. However the biometric data and our own breeding experiments point to a relationship with the *Adansonii*.

In her external markings and general behaviour she closely resembles the Indian bee *Apis indica*. Our evaluations show that she is below average in fecundity, restless and very nervous. In spite of this she cannot be described as bad tempered. If unsuitably crossed with other races she becomes very ill-tempered and has the same tendency to pursue interferers as has the Egyptian race group. And yet, in her native land we could deal with her without smoke or any special protection. She is clearly susceptible to cold. Outside her native habitat the pure Sahariensis has no future except for the creation of new combinations.

When it comes to crosses and combination breeding there is no doubt the Saharan bee has an important part to play, provided the crosses are carefully selected. When suitably crossed this bee shows an extraordinary fecundity and capacity for honey production. During our experiments with crosses of this in the summer of 1964 a Sahara-Buckfast F1 averaged 133 kg per colony against a general average of 36.6 kg. This quite outstanding performance was largely due to the phenomenal strength of the colonies and partly to the vitality, longevity, wing power and industry of this first-cross. The strength of the colonies reached such a pitch that most of the hives in common use would have proved totally inadequate. Crosses of this type will manifest a fabulous comb-building ability. They will draw out foundation to perfection and at superlative speed, which is an essential concomitant of outstanding honey-gathering ability and an absence of swarming. These characteristics go hand in hand.

Further journeys in the Sahara gave us the opportunity of making comparisons between queens from different oases. These comparisons showed clearly that there is no notable difference in the characteristics of bees of one oasis and that of another, apart from the slight variations that make their appearance in every race. The Saharan bee is notable for its great uniformity similar to the Egyptian and Cyprian races. As regards colour the progeny shows a variation similar to that of the *Apis indica*, that is, from very light to a dark brown. Uniformly dark bees do not appear but there is a uniformity in the very light colour. In these further testings we came across no especial susceptibility to disease. In the first experiments we made the bees seemed very prone to Paralysis. As regards fecundity and colony strength we had even better results than in the initial evaluations. We have proved to ourselves that a cross Sahara-Buckfast produces the highest colony strength of any other cross. The true value of the Saharan bee lies in a carefully selected cross and combination breeding. Where selective matings are not possible, no experiments with this race should be attempted.

Notes to the Accompanying Table

It is essential for any objective evaluation and comparison to have definite standards and also to make a series of repeated experiments in different environments and under different conditions of honey flow. As a standard measure for the comparisons listed here I have taken the Buckfast bee. Its characteristics are naturally well known to us. The results of the experiments made in our special environment here are simple statements of fact.

So as not to make these tables too complicated, I have limited the results obtained

Results of the Evaluations in Relation to the Buckfast Strain

The first number in each square denotes the results secured from the pure race in question; the second that of a F_1; the third, where indicated, of a F_2. The respective assessment range from +6 to −6.

Races	Fecundity	Industry	Resistance to brood diseases	Resistance to adult diseases	Disinclination to swarm	Longevity	Wing-power	Hardiness	Keen sense of smell	Readiness to enter supers	Comb-building ability	Good temper	Calm behaviour	Propolization	Absence of brace-comb	Sense of orientation
Buckfast	+4	+4	+3	+5	+6	+2	+2	+5	+5	+6	+6	+6	+5	−5	−5	−1
Ligustica	+3/+4	+2/+3	+3/+4	+3/+3	+3/+1	+1/+2	+1/+2	+1/+2	+3/+3	+4/+4	+4/+4	+4/+5	+3/+4	+2/+1	+1/−1	−2/−1
Carnica	+2/+2	+3/+4	+5/+5	+2/+3	−5/−6	+4/+4	+2/+2	+3/+4	+2/+2	−1/+1	−2/+1	+6/+6	+6/+6	+2/+1	+3/+1	+3/+3
Cecropia	+2/+5	+3/+4	+3/+3	+2/+3	+1/+5	+3/+4	+2/+2	+3/+4	+2/+2	+1/+4	−1/+4	+4/+5	+4/+5	+2/−1	+3/−1	+2/+2
Caucasica	+1/+3	+1/+2	+1/+1	+1/+1	+1/+1	+1/+1	+1/+1	+1/+2	+1/+1	−6/−1	−6/−1	+6/+6	+6/+6	+6/+4	+6/+4	+1/+1
Intermissa	+1/+3 / +4	+4/+5	−4/−4	−3/−1	−4 / −5/+3	+6/+6	+6/+6	+6/+6	+6/+6	+1/+3	+5/+5	−6 / −1/+2	−6/−1	+6/+5	+6/+5	+3/+3
West European races	+1/+3 / +4/+5	+5/+6	−3/−1	−3/−1	−4/+3 / −5/+3	+6/+6	+6/+6	+6/+6	+6/+6	+2/+3	+6/+6	−5 / −1/+2	−5/−2	+6/+4	+6/+4	+3/+3
Mellifica lehzeni	+2/+2	+5/+6	−3/−1	−1/+1	−6/−6	+6/+6	+6/+6	+6/+6	+6/+6	+2/+3	+6/+6	−5/−1	−5/−2	+6/+4	+6/+4	+3/+3
Fasciata	+1/+3	+2/+3	+2/+3	+2/+3	−1 / +2/+3	−1/+1	−6/−5	−6/−1	+3/+4	+1/+4	−1/+3	−5/−1	−5/−1	−6/−4	−6/−4	+6/+6
Cypria	+1/+3/+5	+2/+5	+2/+3	+2/+3	−1 / −4/+3	+2/+3	+2/+3	+3/+5	+4/+5	−1/+3	−1/+3	−5 / −1/+2	−5/−1	+1/−1	−6/−2	+6/+6
Central Anatolica	+1/+3/+5	+6/+6	+2/+3	+3/+4	+2/+5	+6/+6	+6/+6	+5/+5	+3/+4	+1/+3	+2/+2	−1/+2	−1/+2	+3/+2	+3/+2	+3/+3
Sahariensis	+1/+5/+6	+6/+6	+3/+3	+3/+3	+3 / +2/+4	+4/+5	+4/+5	−3/+5	+6/+6	−1/+4	+1/+4	+2 / −1/+2	−6/−1	+2/+1	+3/+4	+4/+3

to the essential races and groups, that is, to those which proved suitable for crosses and for building up new combinations. The fact that some of the crosses only proved of economic value in the F2, and not in the F1 (against the generally held opinion) has been noted where it is of importance.

Only the really important characteristics have been taken into account in these evaluations. The first five form the basis for performance, that is, fertility, industry, resistance to disease of the brood and of the adult bee, and unwillingness to swarm. In the next group are listed longevity, wing power, resistance to weather, sense of smell, storing honey away from the brood, and comb building. These last named characteristics exercise a major influence on the inclination to swarm. The others affect the actual production of the honey. Finally we come to gentleness, steadiness on the combs, propolising, building of brace comb and sense of orientation. These traits are dealt with purely from the management standpoint. They have no influence on honey production. But a good tempered bee, one whose behaviour is quiet, is an essential need for beekeeping today just as is the absence of propolis and brace comb.

The evaluations were measured in twelve grades, but not as 1 to 12, but 6 to 1 and 1 to 6. These were then given a plus or minus sign, which shows the exact gradation. The gradation from one extreme to the other is continuous as shown by the figure and the sign. The only exceptions to this are the figures for propolising and construction of brace comb. To follow the scheme here would give the impression that an increase of these two disadvantageous traits was an advantage. Hence in these two instances $+6$ means the highest amounts of propolis and brace comb and -6 means the lowest.

Putting these characteristics side by side enables one to see at a glance what are the advantages and disadvantages of the different races. At the same time we see what possibilities for crosses and combination are at our disposal.

The Genetic Resources

I have now set out what our experiments and their results have shown, the essential characteristics of the different races known to us at present, and also the breeding possibilities in pure and cross breeding. All the tests were carried out in accordance with some very definite standpoints as the basis because only in this way can we arrive at objective evaluations demanded by a purposeful breeding of the honeybee. As far as possible nothing was left to mere chance. When one works with living creatures one must always be prepared for surprises. There are no findings of universal validity; one can but proceed along general guidelines.

It is very surprising that two new races of bees have been discovered in the past ten years, races about which nothing was previously known. Again thirty years ago we knew next to nothing about the Anatolian group of races. There is no doubt that in the not too distant future further discoveries of this kind will take place. Probably these will be made in the regions south of the Sahara, a habitat of the honeybee which has not been explored up to now.

On the other hand the queen breeder like the plant breeder has to recognise the fact that there is a gradual loss of individual races and local types in the form which was developed by Nature in the distant past. These varieties which for thousands of years possessed the ability to withstand the forces of Nature and the ravages of disease and were clearly gifted with a great number of hereditary characteristics of inestimable value, have today been lost to the modern breeder. It is well known that in the battle against disease it is the wild forms of plants which play a decisive role.

In breeding the honeybee we have to deal only with such wild forms, as there are strictly no 'thoroughbreds'. But many of these 'natural' or geographical races, for example the French local types, which were still about 25 years ago, no longer survive. We have a similar case with the northern Greek variety of the Cecropia. In the first case the loss was due to widespread indiscriminate cross-breeding, and in the second case to a colossal yearly transportation of colonies from all parts of Greece to the northern districts for the honey flow there. Even the Saharan bee in spite of its almost complete isolation in the oases is threatened with extinction. Mistakes have been made with the employment of plant sprays which are injurious to bēes and thus the few colonies left are in danger. Again the modern Carniolan bee has little resemblance in its external markings and its commercial attributes to the Carniolan of former times. The fault here seems to lie in an idealistic approach to breeding.

The greatest danger which today threatens almost every race of the honeybee comes from the prevalent indiscriminate use of mongrel stock at an international level and at the same time from a widespread dissemination of certain very good strains of bees. This of course means that the good characteristics are being propagated, but at the same time it does entail a loss of the genetic riches which were once available.

A realistic approach to breeding cannot be blind to these modern developments. In the years to come it will be increasingly difficult to find genuine representatives of the different geographical races, which are essential for cross-breeding and the formation of new combinations. The beekeeping of the future, to be at all profitable, needs a bee which is good tempered, not given to swarming, and capable of being used for quality and combination breeding joined with the minimal expenditure of time and effort. To preserve and promote these breeding possibilities it is essential to establish reservations to maintain these different races. This maintenance of the races with their original hereditary wealth and individuality is a pre-requisite for any progress in breeding the honeybee.

Conclusion

Apparently our only hope of achieving real progress in the improvement of the genetic composition of the honeybee, open to us, is by way of cross breeding and the synthesization of the wealth of economically valuable traits Nature has placed at our disposal in the different geographical races.

The breeding of pure stock is the indispensable means we have, which enables us to intensify and maintain a particular quality, but we can never thus create a new combination or some new characteristic. By pure breeding we can but intensify and fix what is at hand. But, by the synthesization of new combinations we escape the constrictions pure breeding imposes and can secure positive improvements in the honeybee.

Pure stock undoubtedly provides us with the basic material and groundwork for successful new combinations. Admittedly, we must revert to pure breeding to stabilize and fix the set of new characteristics and to give them permanence, for without this factor all our endeavours would be of little avail. Every new combination must in turn lead step by step to further synthesis of genetic potentialities. Our aim is a progressive, positive and permanent improvement of the honeybee to correspond to the demands of modern beekeeping. In this way alone is a permanent and all-embracing advancement in the improvement of the honeybee possible.

I am well aware that much of what I have written will appear somewhat academic to many beekeepers. However, every well informed beekeeper should be acquainted with the problems at issue in the breeding of the honeybee, even if this knowledge has only a limited application in his own beekeeping. Everyone ought also to realise that our endeavours occupy an exceptional position in this matter of breeding. We have to deal with problems which are completely unknown in all other spheres of similar endeavours.

Glossary

Allele (Allelomorph)	=	An alternative form of a gene occupying a specific locus, or site, on a chromosome. Variation that occurs among individuals of the same species results from the alleles that exist for each trait.
Atavism	=	A reversion to an ancestral type. The Swiss "Nigra" is an instance of this kind.
Bactericides	=	Substances that destroy bacteria.
Biometrics	=	The differentiation of races, based on measurements of their external characteristics.
Breeding	=	This term has many connotations. Here it is restricted to a cumulative permanent improvement of the honeybee on a genetic basis.
Chromosomes	=	The structures in the cell nucleus which carry the genes.
Conglomeration	=	A dissimilar collection of genetic potentialities.
Cross-over	=	A permanent exchange between groups of reciprocal genes.
Cubital index	=	A measurement obtained by measuring certain aspects of wing venation in the Hymenoptera and which enables conclusions to be reached on the possible racial origin of the particular insect.
Diploid	=	Having the chromosomes in pairs.
Dominance	=	The ability of a gene to prevail in the offspring over its opposite or recessive trait.
Erosion	=	A progressive deterioration in stamina or some other trait.
Evolution	=	The gradual development of complex organisms from simpler ones, assumed to be brought about by natural selection (the survival of the fittest) over very long periods of time.
Ecotype	=	A local strain possessing distinctive characteristics evolved in a particular environment over the course of time.
Gene	=	The unit responsible for the material of inheritance.
Genotype	=	The genetic constitution of an individual, as contrasted with its visible characteristics.
Haploid	=	Having only one set of chromosomes.
Hermaphrodite	=	A creature manifesting both male and female characteristics.
Heterosis	=	Hybrid vigour.
Heterozygous	=	Containing both dominant and recessive genes for a given trait. Heterozygous individuals do not breed true to type.
Integument	=	The outer protective covering of the body of an animal.
Inter se	=	Matings within a family or with close relations.
Koersystem	=	A mode of selection confined to external racial characteristics — widely used in breeding the Carniolan bee on the Continent.
Lethal gene	=	A gene which introduces a characteristic which is fatal to the organism.
Mongrel	=	An animal of nondescript pedigree.
Morphology	=	The science which deals with the form and structure of animals and plants.
Mutation	=	The inception of a heritable variation brought about by structural and numerical chromosome changes.

Parthenogenesis	=	The ability to produce offspring from unfertilized eggs.
Phenotype	=	A type determined by visible characteristics which are common to a group and which result from the interaction of environmental factors with the hereditary characteristics.
Polymeral	=	A form of inheritance subject to a series of genetic factors that progressively intensify a particular trait.
Physiological	=	Related to the life activities and behaviour of living things.
Resistance	=	The ability to withstand, to a greater or lesser degree, infection, parasites and disease.
Segregation	=	The separation of characters in the paired chromosomes following a reduction division and which renders a reallocation in the offspring possible.
Septicaemia	=	An infection affecting the blood or lymph.
Synthesization	=	The uniting, in conjunction with cross-breeding, of the desired characteristics of two or more races into a new fixed genetic combination and stable uniform offspring.

Symbols

F1	=	A first filial generation of a cross.
P	=	Parents.
P1	=	The first parental generation.
♀	=	A queen bee.
♂	=	A drone.
⚥	=	A worker bee.

Lightning Source UK Ltd.
Milton Keynes UK
UKHW051906301120
374378UK00008B/1206